Titles in This Series

760 **Markus Banagl,** Extending intersection homology type invariants to non-Witt spaces, 2002

759 **Donald M. Davis,** From representation theory to homotopy groups, 2002

758 **Alan Forrest, John Hunton, and Johannes Kellendonk,** Topological invariants for projection method patterns, 2002

757 **Douglas Bowman,** q-difference operators, orthogonal polynomials, and symmetric expansions, 2002

756 **José Ignacio Cogolludo-Agustín,** Topological invariants of the complement to arrangements of rational plane curves, 2002

755 **M. A. Mandell and J. P. May,** Equivariant orthogonal spectra and S-modules, 2002

754 **Edward L. Green, Idun Reiten, and Øyvind Solberg,** Dualities on generalized Koszul algebras, 2002

753 **Daniel Panazzolo,** Desingularization of nilpotent singularities in families of planar vector fields, 2002

752 **Linus Kramer,** Homogeneous spaces, Tits buildings, and isoparametric hypersurfaces, 2002

751 **Bruce Allison, Georgia Benkart, and Yun Gao,** Lie algebras graded by the root systems BC_r, $r \geq 2$, 2002

750 **Masaki Izumi and Hideki Kosaki,** Kac algebras arising from composition of subfactors: General theory and classification, 2002

749 **Nanhua Xi,** The based ring of two-sided cells of affine Weyl groups of type $\widetilde{A}_{n-1}$, 2002

748 **Jürgen Ritter and Alfred Weiss,** The lifted root number conjecture and Iwasawa theory, 2002

747 **Armand Borel, Robert Friedman, and John W. Morgan,** Almost commuting elements in compact Lie groups, 2002

746 **Peter Niemann,** Some generalized Kac-Moody algebras with known root multiplicities, 2002

745 **Mikhail A. Lifshits and Werner Linde,** Approximation and entropy numbers of Volterra operators with application to Brownian motion, 2002

744 **Roger Chalkley,** Basic global relative invariants for homogeneous linear differential equations, 2002

743 **Heng Sun,** Spectral decomposition of a covering of $GL(r)$: the Borel case, 2002

742 **J. E. Gilbert, Y. S. Han, J. A. Hogan, J. D. Lakey, D. Weiland, and G. Weiss,** Smooth molecular functions and singular integral operators, 2002

741 **Francisco Santos,** Triangulations of oriented matroids, 2002

740 **Rick Durrett,** Mutual invadability implies coexistence in spatial models, 2002

739 **Georgios K. Alexopoulos,** Sub-Laplacians with drift on Lie groups of polynomial volume growth, 2002

738 **Yasuro Gon,** Generalized Whittaker functions on $SU(2,2)$ with respect to the Siegel parabolic subgroup, 2002

737 **Arjen Doelman, Robert A. Gardner, and Tasso J. Kaper,** A stability index analysis of 1-D patterns of the Gray-Scott model, 2002

736 **Wojciech Chachólski and Jérôme Scherer,** Homotopy theory of diagrams, 2002

735 **Martina Brück, Xi Du, Joonsang Park, and Chuu-Lian Terng,** The submanifold geometries associated to Grassmannian systems, 2002

734 **Michel Van den Bergh,** Blowing up of non-commutative smooth surfaces, 2001

733 **Milé Krajčevski,** Tilings of the plane, hyperbolic groups and small cancellation conditions, 2001

Extending Intersection Homology Type Invariants to Non-Witt Spaces

MEMOIRS
of the
American Mathematical Society

Number 760

Extending Intersection Homology Type Invariants to Non-Witt Spaces

Markus Banagl

November 2002 • Volume 160 • Number 760 (second of 5 numbers) • ISSN 0065-9266

American Mathematical Society
Providence, Rhode Island

2000 *Mathematics Subject Classification.*
Primary 55N33, 57N80, 57R20, 32S60; Secondary 32S45, 57Q50.

Library of Congress Cataloging-in-Publication Data

Banagl, Markus, 1971–
Extending intersection homology type invariants to non-Witt spaces / Markus Banagl.
p. cm. — (Memoirs of the American Mathematical Society, ISSN 0065-9266 ; no. 760)
Includes bibliographical references.
ISBN 0-8218-2988-2 (alk. paper)
1. Intersection homology theory. 2. Duality theory (Mathematics) I. Title. II. Series.

QA612.32.B36 no. 760
[QA3.A54]
510 s—dc21
[514′.23] 2002074586

Memoirs of the American Mathematical Society

This journal is devoted entirely to research in pure and applied mathematics.

Subscription information. The 2003 subscription begins with volume 161 and consists of six mailings, each containing one or more numbers. Subscription prices for 2003 are $555 list, $444 institutional member. A late charge of 10% of the subscription price will be imposed on orders received from nonmembers after January 1 of the subscription year. Subscribers outside the United States and India must pay a postage surcharge of $31; subscribers in India must pay a postage surcharge of $43. Expedited delivery to destinations in North America $35; elsewhere $130. Each number may be ordered separately; *please specify number* when ordering an individual number. For prices and titles of recently released numbers, see the New Publications sections of the *Notices of the American Mathematical Society.*

Back number information. For back issues see the *AMS Catalog of Publications.*

Subscriptions and orders should be addressed to the American Mathematical Society, P. O. Box 845904, Boston, MA 02284-5904, USA. *All orders must be accompanied by payment.* Other correspondence should be addressed to 201 Charles Street, Providence, RI 02904-2294, USA.

Memoirs of the American Mathematical Society is published bimonthly (each volume consisting usually of more than one number) by the American Mathematical Society at 201 Charles Street, Providence, RI 02904-2294, USA. Periodicals postage paid at Providence, RI. Postmaster: Send address changes to Memoirs, American Mathematical Society, 201 Charles Street, Providence, RI 02904-2294, USA.

This publication is indexed in *Science Citation Index*®, *SciSearch*®, *Research Alert*®, *CompuMath Citation Index*®, *Current Contents*®/*Physical, Chemical & Earth Sciences.*
Printed in the United States of America.

∞ The paper used in this book is acid-free and falls within the guidelines established to ensure permanence and durability.
Visit the AMS home page at http://www.ams.org/

10 9 8 7 6 5 4 3 2 1 07 06 05 04 03 02

Contents

Abstract

Intersection homology theory provides a way to obtain generalized Poincaré duality, as well as a signature and characteristic classes, for singular spaces. For this to work, one has had to assume however that the space satisfies the so-called Witt condition. We extend this approach to constructing invariants to spaces more general than Witt spaces.

We present an algebraic framework for extending generalized Poincaré duality and intersection homology to singular spaces X not necessarily Witt. The initial step in this program is to define the category $SD(X)$ of complexes of sheaves suitable for studying intersection homology type invariants on non-Witt spaces. The objects in this category can be shown to be the closest possible self-dual "approximation" to intersection homology sheaves. It is therefore desirable to understand the structure of such self-dual sheaves and to isolate the minimal data necessary to construct them. As the main tool in this analysis we introduce the notion of a Lagrangian structure (related to the familiar notion of Lagrangian submodules for $(-1)^k$-Hermitian forms, as in surgery theory). We demonstrate that every complex in $SD(X)$ has naturally associated Lagrangian structures and conversely, that Lagrangian structures serve as the natural building blocks for objects in $SD(X)$. Our main result asserts that there is in fact an equivalence of categories between $SD(X)$ and a twisted product of categories of Lagrangian structures. This may be viewed as a Postnikov system for $SD(X)$ whose fibers are categories of Lagrangian structures.

The question arises as to which varieties possess Lagrangian structures. To begin to answer that, we define the model-class of varieties with an ordered resolution and use block bundles to describe the geometry of such spaces. Our main result concerning these is that they have associated preferred Lagrangian structures, and hence self-dual generalized intersection homology sheaves.

Received by the editor March 8, 2001.

2000 *Mathematics Subject Classification.* Primary 55N33, 57N80, 57R20, 32S60; Secondary 32S45, 57Q50.

Key words and phrases. Geometric topology, singular spaces, intersection homology, characteristic classes, perverse sheaves, Poincaré duality, bordism theory, resolution of singularities.

The author was supported in part by NSF grant DMS-0072550.

CHAPTER 1

Introduction

1. History

The question of extending Poincaré duality to a suitable kind of homology on spaces more general than manifolds was attacked by Goresky and MacPherson, using sheaf-theoretic methods, in their fundamental paper [**GM83**]. They construct a differential complex of sheaves $\mathbf{IC}^\bullet_{\bar{p}}(X)$ defined on a pseudomanifold X and depending on a parameter $\bar{p}$, the so-called perversity. The hypercohomology groups of this complex are the intersection homology groups $IH^{\bar{p}}_*$. Moreover, the authors define the intersection pairing $IH^{\bar{p}}_*(X) \otimes IH^{\bar{q}}_*(X) \to H_*(X)$, where $\bar{p}$ and $\bar{q}$ are complementary perversities. The pairing results from the duality isomorphism (in the derived category of differential complexes of sheaves) $\mathbf{IC}^\bullet_{\bar{p}} \cong \mathcal{D}(\mathbf{IC}^\bullet_{\bar{q}})[\dim X]$, where $\mathcal{D}$ denotes the Borel-Moore-Verdier dualizing functor. Since perversities are integer valued, there are two "middle perversities": the lower middle ($\bar{m}$) and the upper middle ($\bar{n}$) perversity. If X has only strata of even codimension, then $\mathbf{IC}^\bullet_{\bar{m}} = \mathbf{IC}^\bullet_{\bar{n}}$ as $\bar{m}(2k) = \bar{n}(2k)$, and hence $\mathbf{IC}^\bullet_{\bar{m}} \cong \mathcal{D}(\mathbf{IC}^\bullet_{\bar{m}})[\dim X]$, i.e. $\mathbf{IC}^\bullet_{\bar{m}}$ is self-dual. Therefore one obtains an adequate generalization of Poincaré duality to such spaces. In [**Sie83**], Siegel has extended this approach to Witt spaces. A space X is Witt, if the middle dimensional, lower middle perversity intersection homology of the links of all odd codimensional strata vanishes. In that case, the canonical morphism $\mathbf{IC}^\bullet_{\bar{m}} \to \mathbf{IC}^\bullet_{\bar{n}}$ is a quasi-isomorphism and again $\mathbf{IC}^\bullet_{\bar{m}}(X)$ is self-dual. In particular, one obtains (provided $\dim X$ is even) a signature $\sigma(X)$ of X, and following the Thom-Pontrjagin-Milnor program, or using the methods of [**CSW91**], L-classes.

2. Motivation

Here we develop methods to extend the above approach to obtaining generalized Poincaré duality and producing invariants for pseudomanifolds to spaces more general than Witt spaces. For a number of potential purposes, requiring a pseudomanifold to be Witt is too restrictive: Certain compactifications (see section 6.1 of this introduction), quotients (see section 5), or real algebraic varieties of interest may be stratified pseudomanifolds, but may fail to be stratifiable so as to satisfy the Witt space condition.

EXAMPLE 1.1. The open cone $T \times \mathbb{R}/(T \times \{0\})$ on a real 2-torus T is a real algebraic variety X^3 which is a pseudomanifold with one isolated singularity of odd codimension, the cone-point. The morphism $\mathbf{IC}^\bullet_{\bar{m}}(X) \to \mathbf{IC}^\bullet_{\bar{n}}(X)$ is not a quasi-isomorphism as $H^1(T) \neq 0$ (i.e. X is not Witt). Thus $\mathbf{IC}^\bullet_{\bar{m}}(X)$ is not self-dual. The theory presented in this paper allows us to construct a self-dual perverse sheaf on X, whose self-duality isomorphism is compatible (in an appropriate sense, cf. section 3) with the classical intersection pairing.

On the other hand, consideration of examples such as the cone on complex projective space $\mathbb{C}P^2$, shows that an attempt to carry out this program will not be meaningful on the class of all pseudomanifolds. However, note that in this example, the link of the singularity has nonzero signature. This observation suggests that one should look at spaces having the property that the links of all singular strata of odd codimension have signature zero. If a space X has just one singular stratum, the links are all contained in the top stratum of X, and we compute their signature with respect to the usual cup-product pairing on middle-dimensional cohomology. If we assume these signatures to be zero, the middle dimensional cohomology of the links will contain a Lagrangian[1] subspace (which, of course, is not unique). The idea is to make a "continuous" choice of Lagrangian subspaces along the singular stratum, and to use this data in constructing a self-dual sheaf $\mathbf{IC}^\bullet_{\mathcal{L}}$ that "interpolates" between $\mathbf{IC}^\bullet_{\bar{m}}$ and $\mathbf{IC}^\bullet_{\bar{n}}$, more precisely: Construct $\mathbf{IC}^\bullet_{\mathcal{L}}$ and morphisms $\mathbf{IC}^\bullet_{\bar{m}} \to \mathbf{IC}^\bullet_{\mathcal{L}} \to \mathbf{IC}^\bullet_{\bar{n}}$ such that $\mathbf{IC}^\bullet_{\mathcal{L}} \cong \mathcal{D}\mathbf{IC}^\bullet_{\mathcal{L}}[\dim X]$ and this isomorphism commutes with $\mathbf{IC}^\bullet_{\bar{m}} \cong \mathcal{D}\mathbf{IC}^\bullet_{\bar{n}}[\dim X]$.

3. The Main Result: A Postnikov System of Lagrangian Structures

We start out by defining a category $SD(X)$ (definition 2.2) associated to a pseudomanifold X that captures the satisfactory candidate sheaves for extending intersection homology self-dually to non-Witt spaces. $SD(X)$ is a full subcategory of the derived category of constructible bounded complexes of sheaves on X, whose objects $\mathbf{S}^\bullet$ satisfy a set of four axioms: Normalization over the top stratum, a lower bound requirement, an upper middle perversity stalk condition, and the requirement that $\mathbf{S}^\bullet$ have an associated self-duality isomorphism $d : \mathcal{D}\mathbf{S}^\bullet[n] \cong \mathbf{S}^\bullet$ (here $n = \dim X$). Note that the first three axioms are identical with the corresponding conditions in the Goresky-MacPherson axiomatics [**GM83**] for the intersection chain sheaf $\mathbf{IC}^\bullet_{\bar{n}}(X)$, $\bar{n}$ = upper middle perversity. If X is Witt, for example a complex algebraic variety, then $SD(X)$ contains, up to isomorphism, only one object $\mathbf{IC}^\bullet_{\bar{m}} \cong \mathbf{IC}^\bullet_{\bar{n}}$. If X is not Witt, for example a real algebraic variety whose links carry middle dimensional intersection homology, then $SD(X)$ may contain nonisomorphic objects or may be empty. The defining axioms for $SD(X)$ imply that this is a full subcategory of the category of perverse sheaves on X.

What is the value of $SD(X)$? The answer is provided by the following two facts: On the one hand, every object $\mathbf{S}^\bullet$ in $SD(X)$ is self-dual and hence gives rise to a generalized Poincaré duality for its associated hypercohomology groups, on the other hand one can show (theorem 2.2) that $\mathbf{S}^\bullet$ "interpolates" in a rather precise way between $\mathbf{IC}^\bullet_{\bar{m}}(X)$ and $\mathbf{IC}^\bullet_{\bar{n}}(X)$, the lower/upper middle intersection chain complexes. Concretely, there exist unique morphisms

$$\mathbf{IC}^\bullet_{\bar{m}}(X) \xrightarrow{\alpha} \mathbf{S}^\bullet \xrightarrow{\beta} \mathbf{IC}^\bullet_{\bar{n}}(X)$$

[1] In our use of the term "Lagrangian subspace" we follow Andrew Ranicki's second edition of Wall's book [**Wal99**], e.g. lemma 5.3 or §17G. There, a free submodule H of a free Λ-module G with a $(-1)^k$-Hermitian form $\lambda : G \times G \to \Lambda$ (where Λ is a ring with involution) is called *Lagrangian* if $\lambda(H \times H) = 0$ and the map $G/H \to \mathrm{Hom}_\Lambda(H, \Lambda)$ induced by λ is an isomorphism. Thus for us, as in [**Wal99**], "Lagrangian" is equivalent to "maximally isotropic."

compatible with the top stratum normalizations, factoring the canonical $\mathbf{IC}^\bullet_{\bar{m}} \to \mathbf{IC}^\bullet_{\bar{n}}$, and such that

$$\begin{array}{ccc} \mathbf{IC}^\bullet_{\bar{m}}(X) & \xrightarrow{\alpha} & \mathbf{S}^\bullet \\ \uparrow \cong & & \cong \uparrow d \\ \mathcal{D}\mathbf{IC}^\bullet_{\bar{n}}(X)[n] & \xrightarrow{\mathcal{D}\beta[n]} & \mathcal{D}\mathbf{S}^\bullet[n] \end{array}$$

commutes (the left vertical isomorphism is the Goresky-MacPherson duality isomorphism). This demonstrates that the objects of $SD(X)$ are the closest self-dual "approximation" to intersection homology possible on non-Witt spaces.

In light of the preceding discussion, it is desirable to understand the structure of $SD(X)$. We have conducted such an analysis: it culminates in a certain Postnikov system, theorem 2.10 below. The main tool and indeed the building blocks of complexes in $SD(X)$ are what we call *Lagrangian structures*. To define this notion, consider a pure stratum $V \subset X$ of codimension k, let $U = X - \overline{V}$ be the complement of the closure of V, and $i : U \hookrightarrow U \cup V$ the inclusion. If $\mathbf{S}^\bullet$ is a self-dual sheaf in $SD(U)$, the restricted derived sheaf $\mathbf{H}^s(Ri_*\mathbf{S}^\bullet)|_V$ in dimension $s = \bar{n}(k) - n$ of the direct image will be equipped with a nonsingular bilinear pairing

$$\mathbf{H}^s(Ri_*\mathbf{S}^\bullet)|_V \otimes \mathbf{H}^s(Ri_*\mathbf{S}^\bullet)|_V \longrightarrow \mathbb{R}_V. \tag{1}$$

A sheaf monomorphism $\mathcal{L} \to \mathbf{H}^s(Ri_*\mathbf{S}^\bullet)|_V$, $\mathcal{L} \in Sh(V)$, is a *Lagrangian structure*, if every stalk $\mathcal{L}_x$, $x \in V$, is a Lagrangian subspace of $\mathbf{H}^s(Ri_*\mathbf{S}^\bullet)_x$ with respect to the pairing. It is advantageous to use an equivalent definition which is more adapted to the triangulated structure of the derived category and to Verdier duality. So the definition that we actually work with (definition 2.3) is phrased in terms of an algebraic notion of nullcobordism of sheaves as introduced in [**CS91**]. Lagrangian structures can be conveniently organized into a morphism category $\operatorname{Lag}(V)$.

Lagrangian structures are the key to understanding $SD(X)$. Their relevance is explained by the following two constructions:

1. Extracting Lagrangian structures from self-dual sheaves (chapter 2, section 4): There exists a covariant functor

$$\Lambda : SD(U \cup V) \longrightarrow \operatorname{Lag}(V).$$

   This means that every self-dual perverse sheaf has naturally associated Lagrangian structures.
2. Lagrangian structures as building blocks for self-dual sheaves (chapter 2, section 5): Given $\mathbf{S}^\bullet \in SD(U)$, a Lagrangian structure $\mathcal{L}$ along the adjacent lower dimensional pure stratum V naturally gives rise to a self-dual sheaf on the union $U \cup V$. Formally, one has a covariant functor[2]

$$\begin{array}{rccc} \boxplus : & SD(U) \rtimes \operatorname{Lag}(V) & \longrightarrow & SD(U \cup V), \\ & (\mathbf{S}^\bullet, \mathcal{L}) & \mapsto & \mathbf{S}^\bullet \boxplus \mathcal{L} \end{array}$$

[2] $SD(U) \rtimes \operatorname{Lag}(V)$ denotes a fibered category over $SD(U)$ (see chapter 2, section 5.2). This is necessitated by a Lagrangian structure being a map into the derived sheaf of a given complex of sheaves.

One can show that the functors

$$SD(U) \rtimes \mathrm{Lag}(V) \underset{(i^*,\Lambda)}{\overset{\boxplus}{\rightleftarrows}} SD(U \cup V)$$

induce an equivalence of categories (i^* denotes restriction to U). Using this fact, we obtain our main result concerning the algebraic structure of the category $SD(X)$ of self-dual sheaves compatible with intersection homology theory: a Postnikov system whose fibers are categories of Lagrangian structures along the various pure singular strata (write $U_k = X - X_{n-k}$):

THEOREM. 2.10. *Let $n = \dim X$ be even. There is an equivalence of categories*

$$SD(X) \simeq \mathrm{Lag}(U_n - U_{n-1}) \rtimes \mathrm{Lag}(U_{n-2} - U_{n-3}) \rtimes \ldots \rtimes$$

$$\mathrm{Lag}(U_4 - U_3) \rtimes \mathrm{Const}(U_2).$$

(Here $\mathrm{Const}(U_2)$ denotes the category whose single object is the constant sheaf $\mathbb{R}_{U_2}$ on the top stratum U_2 and whose morphisms are sheaf maps $\mathbb{R}_{U_2} \to \mathbb{R}_{U_2}$.) The theorem is phrased for even-dimensional spaces owing to the choice of sign in axiom (**SD4**) of definition 2.2. The appropriate category $SD^o(X)$ for odd-dimensional X is obtained by changing (**SD4**) to $\mathcal{D}d[n] = -d$, and the analog of theorem 2.10 for $SD^o(X)$ holds. Sign issues are briefly discussed in section 8 of this introduction, whereas a detailed technical account is provided in appendix A.

The idea of employing Lagrangian subspaces in order to obtain self-duality is present in an L^2-cohomology setting as J. Cheeger's "$*$-invariant boundary conditions," see [**Che79**], [**Che80**] and [**Che83**], and is also invoked in unpublished work of J. Morgan on the characteristic variety theorem.

4. Consequences: Characteristic Classes

Let X be a closed oriented non-Witt space of even dimension n. If Lagrangian structures $\mathcal{L}$ exist along all odd-codimensional strata, then theorem 2.10 tells us that X carries a perverse sheaf $\mathbf{IC}^\bullet_{\mathcal{L}}(X)$ with a self-duality isomorphism $d : \mathcal{D}\mathbf{IC}^\bullet_{\mathcal{L}}(X)[n] \xrightarrow{\cong} \mathbf{IC}^\bullet_{\mathcal{L}}(X)$. By Borel-Moore duality,

$$\mathcal{H}^{-k}(X; \mathcal{D}\mathbf{IC}^\bullet_{\mathcal{L}}(X)[n]) \cong \mathrm{Hom}(\mathcal{H}^{-k}(X; \mathbf{IC}^\bullet_{\mathcal{L}}(X)), \mathbb{R})$$

(where $2k = n$), whence d induces a nonsingular pairing

$$IH^{\mathcal{L}}_k(X) \otimes IH^{\mathcal{L}}_k(X) \longrightarrow \mathbb{R}, \tag{2}$$

where we set $IH^{\mathcal{L}}_k(X) = \mathcal{H}^{-k}(X; \mathbf{IC}^\bullet_{\mathcal{L}}(X))$. Thus, denoting the signature of (2) by $\sigma(X)$, we obtain a signature for non-Witt spaces.

To define L-classes, we recall the existence and uniqueness result on L-classes of self-dual complexes of sheaves:

THEOREM 1.1 ([**CS91**]). *Let $\boldsymbol{S}^\bullet \in D^b(X)$ be a self-dual complex of sheaves. There exist unique classes $L_k(\boldsymbol{S}^\bullet) \in H_k(X; \mathbb{Q})$ with the following properties:*

(i) $\epsilon_* L_0(\boldsymbol{S}^\bullet) = \sigma(\boldsymbol{S}^\bullet)$

(ii) *If $j : Y^m \hookrightarrow X^n$ is a normally nonsingular inclusion with trivial normal bundle, then*

$$L_{k-n+m}(j^! \boldsymbol{S}^\bullet) = j^! L_k(\boldsymbol{S}^\bullet),$$

where $j^!$ on the right hand side is the Thom map $j^! : H_k(X) \longrightarrow H_{k-n+m}(Y)$.

We define the L-class of X by letting

$$L_i(X) = L_i(\mathbf{IC}^\bullet_{\mathcal{L}}(X)).$$

5. Ordered Resolutions — A Model Construction

The question arises as to which varieties have Lagrangian structures. We describe a model-class of such spaces in chapter 3: A PL-variety X (in the sense of [**Sto72**]) has an *ordered resolution*, if it can be obtained from a closed oriented PL-manifold M as follows: Embed a closed manifold $E \hookrightarrow M$, which is the total space of a fiber bundle over a manifold base B. Define the space X_1 as the fiberwise quotient of M with respect to the fibration. X_1 is a PL-variety with one singular stratum isomorphic to B. Assume inductively that X_k has been constructed. Embed the total space of a fiber bundle (base and fibers manifolds) in the bottom stratum of X_k. Let X_{k+1} be the fiberwise quotient of X_k. The resolution map is the composition of the quotient maps. Our result on varieties with ordered resolutions is the following:

THEOREM. 5.1. *Let X^n be a PL-variety which has an ordered resolution $M^n \to X$, where M is a closed oriented PL-manifold. Then there exist a complex of sheaves $\boldsymbol{IC}^\bullet_{\mathcal{L}}(X) \in D^b(X)$, morphisms*

$$\boldsymbol{IC}^\bullet_{\bar{m}}(X) \to \boldsymbol{IC}^\bullet_{\mathcal{L}}(X) \to \boldsymbol{IC}^\bullet_{\bar{n}}(X)$$

which factor the canonical morphism $\boldsymbol{IC}^\bullet_{\bar{m}}(X) \to \boldsymbol{IC}^\bullet_{\bar{n}}(X)$, and a self-duality isomorphism

$$\mathcal{D}\boldsymbol{IC}^\bullet_{\mathcal{L}}(X)[n] \cong \boldsymbol{IC}^\bullet_{\mathcal{L}}(X)$$

such that the following diagram commutes:

$$\begin{array}{ccc} \boldsymbol{IC}^\bullet_{\mathcal{L}}(X) & \longrightarrow & \boldsymbol{IC}^\bullet_{\bar{n}}(X) \\ \uparrow \cong & & \uparrow \cong \\ \mathcal{D}\boldsymbol{IC}^\bullet_{\mathcal{L}}(X)[n] & \longrightarrow & \mathcal{D}\boldsymbol{IC}^\bullet_{\bar{m}}(X)[n] \end{array}$$

A few remarks on this theorem: The most intriguing aspect of ordered resolutions is that the Lagrangian structures necessary for constructing $\mathbf{IC}^\bullet_{\mathcal{L}}$ are naturally associated to the resolution, so that no arbitrary choices have to be made. We work in the PL-category as this enables us to use block bundle theory as developed by Rourke and Sanderson [**RS68**] (manifold case) and cone block bundle theory as developed by Stone [**Sto72**] (singular case). These tools allow us to describe the geometry of spaces with ordered resolutions (e.g. what are the links of strata?), which feeds into the sheaf theoretic constructions. We believe however that many aspects of these geometric ideas could probably be phrased more elegantly in terms of the homotopically stratified sets of [**Qui88**]. The proof of theorem 5.1 proceeds inductively, following the resolution map. At each stage a Lagrangian structure is constructed, followed by an application of theorem 2.10, building up the self-dual sheaf. The key is to obtain Lagrangian subspaces in the hypercohomology of the links with coefficients in the already available sheaf, which then assemble to give the Lagrangian structure. To organize this efficiently, we introduce in chapter 4 a cobordism group Ω_*^{SD}, whose elements are equivalence classes of closed pseudomanifolds supporting a self-dual perverse sheaf. In particular, an appropriate notion of cobordism and boundary operator for such objects can be defined. A pair (pseudomanifold, self-dual sheaf) has a tautological signature associated to it. We prove that this signature is a cobordism invariant, theorem 4.1 and its two corollaries 4.1 and 4.2. Furthermore, a map $\Omega_*^{\text{Witt}} \longrightarrow \Omega_*^{SD}$ is defined and the composition $\Omega_*^{\text{Witt}} \longrightarrow \Omega_*^{SD} \longrightarrow \mathbb{Z}$ is a factorization of the signature $\Omega_*^{\text{Witt}} \longrightarrow \mathbb{Z}$. The group Ω_*^{SD} (and its generalized version which allows for twisted coefficient systems on the top stratum, see section 9), is of independent interest and has been used e.g. in [**BC00**].

Note that in general, the pushforward under the resolution map of the intersection chain complex on the manifold is not quasi-isomorphic to the complex $\mathbf{IC}^\bullet_{\mathcal{L}}$ constructed in the theorem. The pushforward will be self-dual since the map is proper, but it will not necessarily satisfy the required stalk conditions. Ordered resolutions are not assumed to be small maps.

6. Applications

6.1. Hilbert Modular Surfaces. We sketch a major application of the theory to compactifications of Hilbert modular surfaces. Details will appear in the forthcoming manuscript [**BK01**]. Let K be a totally real algebraic number field and $\mathcal{O}_K$ the ring of algebraic integers in K. Consider the Hilbert modular surface $X = (H \times H)/\Gamma$, where H is the upper half plane and $\Gamma = SL_2(\mathcal{O}_K)/\{\pm 1\}$ the Hilbert modular group. This complex surface is not compact, and various compactifications have been studied. The reductive Borel-Serre compactification has the advantage that the Hecke operators extend to this compactification (which is not true for the other compactifications). We may represent the Hilbert modular surface as

$$X = \Gamma \backslash G(\mathbb{R})/K,$$

with G the algebraic group $G = SO^\circ(q)$ (q an appropriate quadratic form), $G(\mathbb{R})$ the real points of G, K the maximal compact subgroup. The reductive Borel-Serre compactification $\bar{X}$ adds a stratum X_P for each Γ-conjugacy class of parabolic subgroups P of G. We show that every X_P is topologically a circle. Following the paper [**GHM94**], we can describe the link-bundle of these strata as well: The link L turns out to be a 2-torus. In particular, $\bar{X}$ is a real 4-dimensional pseudomanifold which is not a Witt space. There exists a Lagrangian subspace in $H^1(L)$. Furthermore, we have investigated the monodromy as one goes around the link-bundle over the circle, and it turns out that there exists a Lagrangian structure along the circle. Thus, theorem 2.10 applies, and we can construct self-dual perverse sheaves on $\bar{X}$. In particular, we obtain a signature and L-classes for the reductive Borel-Serre compactification of Hilbert modular surfaces. The author wishes to thank R. MacPherson for bringing this problem to his attention.

6.2. Spaces Without Poincaré Duality. Conversely, the Postnikov system can be applied in showing that a given space cannot satisfy Poincaré duality in any theory compatible with intersection homology. Consider, say, the product of a circle with the suspension of two dimensional complex projective space $X^6 = S^1 \times \Sigma\mathbb{C}P^2$. The set of singularities consists of two disjoint circles, $S^1 \times \{c_+\} \sqcup S^1 \times \{c_-\}$, where $c_\pm$ are the two cone-points of $\Sigma\mathbb{C}P^2$. Suppose there were a self-dual complex of sheaves on X, compatible with the intersection chain sheaves, so that the corresponding hypercohomology groups satisfy Poincaré duality. Such a sheaf would define an object in $SD(S^1 \times \Sigma\mathbb{C}P^2)$. The Postnikov system implies the existence of a Lagrangian structure at the singularities. But the link is $\mathbb{C}P^2$, which has second Betti number $b_2 = 1$. Thus, there is no Lagrangian subspace in the middle dimensional cohomology of the link. Therefore, X cannot satisfy Poincaré duality.

7. Further Developments

As discussed above, intersection homology on varieties that do not satisfy the Witt space condition depends on choosing Lagrangian structures. Since these are not unique, it is natural to investigate the following question: If we vary the Lagrangian structure, how does the signature of the resulting self-dual sheaf vary? How do the L-classes vary? Some recent preliminary evidence suggests that in certain interesting situations the characteristic classes are indeed independent of the choice of Lagrangian structure.

8. Sign Questions

Some comments on the sign conventions used:

- In axiom (**SD4**) of definition 2.2, we require the self-duality isomorphism $d : \mathcal{D}\mathbf{L}^\bullet[n] \cong \mathbf{L}^\bullet$ of an object $\mathbf{L}^\bullet \in SD(X^n)$ to satisfy $\mathcal{D}d[n] = +d$. In the presence of the normalization (**SD1**), this is the correct choice of sign for a space X of *even* dimension. To work on odd-dimensional spaces, one introduces a category $SD^o(X)$ by changing axiom (**SD4**) to $\mathcal{D}d[n] = -d$ and adapting the rest of the theory *mutatis mutandis.*
- For applications it is then important to know whether the bilinear pairing (1) is symmetric or anti-symmetric, since for an anti-symmetric pairing the existence of a Lagrangian subspace is automatic (although monodromy issues present a "secondary obstruction" for the existence of an actual Lagrangian structure). The answer is as follows: (1) is symmetric when the dimension of the links of the components of V is congruent 0 mod 4, and anti-symmetric when that dimension is congruent 2 mod 4.

The above statements are proven in appendix A.

9. Some Remarks on Coefficients

Throughout this paper, we use real coefficients. This is not essential, one could work over any field $\mathbb{k}$ of characteristic $\neq 2$. Of course for general $\mathbb{k}$ (e.g. $\mathbb{k} = \mathbb{C}$ the complex numbers), the signature of a bilinear ± 1-symmetric pairing is not available, but all we need to construct self-dual sheaves is the notion of a Lagrangian subspace for such a pairing on a $\mathbb{k}$-vector space. The use of field coefficients, however, is essential. One reason is related to Borel-Moore-Verdier duality (we need identifications such as $\mathcal{H}^k(X; \mathcal{D}\mathbf{A}^\bullet) \cong \operatorname{Hom}(\mathcal{H}^{-k}(X; \mathbf{A}^\bullet), \mathbb{k})$), another is related to the very definition of a self-dual sheaf: If $\mathbb{k}$ is a field of characteristic $\neq 2$, we may define self-duality as the existence of a quasi-isomorphism $d : \mathbf{A}^\bullet \to \mathcal{D}\mathbf{A}^\bullet[n]$ together with a homotopy from d to $\pm\mathcal{D}d[n]$. For more general coefficient rings, one would have to provide higher coherencies (a homotopy from the previous homotopy to its dual, etc.), see [**Wei94**].

Further obvious modifications of the axiomatics given include the use of twisted coefficients on the top stratum. For simplicity, we normalize objects of $SD(X)$ to be the constant sheaf of rank one on the top stratum, cf. definition 2.2, axiom (**SD1**). Replacing (**SD1**) by a relaxed version permitting arbitrary local systems $\mathcal{S}$ (over $\mathbb{k}$, of finite rank) with a nonsingular pairing $\mathcal{S} \otimes \mathcal{S} \longrightarrow \mathbb{k}$, one obtains a more general category $\widetilde{SD}(X)$ for which theorem 2.10 is valid (with $\mathrm{Const}(U_2)$ changed to allow for twisting). For the associated cobordism group $\Omega_*^{\widetilde{SD}}$ (chapter 4) corollary 4.1 remains true.

10. Acknowledgments

The research presented here includes the results of the author's thesis [**Ban99**], written under Professor Sylvain Cappell at the Courant Institute of New York University. I would like to thank my advisor for suggesting this project and for his invaluable guidance. Furthermore I am grateful to Professors Fedor Bogomolov, Jeff Cheeger, Julius Shaneson and Shmuel Weinberger for their encouragement and several helpful discussions.
Most diagrams have been typeset using Paul Taylor's commutative diagram package.

11. Notation

In our notation and conventions we follow [**GM83**], and this includes in particular the indexing scheme. We recall some of that notation: $\mathbf{A}, \mathbf{B}, \mathbf{C}, \ldots$ denote sheaves, $\mathbf{A}^\bullet, \mathbf{B}^\bullet, \mathbf{C}^\bullet, \ldots$ denote complexes of sheaves. The derived sheaf (or cohomology sheaf) in dimension k of a complex of sheaves $\mathbf{A}^\bullet$ is written as $\mathbf{H}^k(\mathbf{A}^\bullet)$. We denote the hypercohomology in dimension k on a space X with coefficients in the complex of sheaves $\mathbf{A}^\bullet$ by $\mathcal{H}^k(X; \mathbf{A}^\bullet)$, whereas $H^k(X; \mathbf{A})$ is sheaf-cohomology of X with coefficient-sheaf $\mathbf{A}$. The Borel-Moore-Verdier dualizing functor is denoted by $\mathcal{D}$ and $\mathbb{D}_X^\bullet$ is the dualizing complex on a space X.

CHAPTER 2

The Algebraic Framework

1. The Lifting Obstruction

Let X^n be an n-dimensional pseudomanifold with a fixed stratification $X = X_n \supset X_{n-2} \supset X_{n-3} \supset \ldots \supset X_0 \supset X_{-1} = \varnothing$ such that X_j is closed in X and $X_j - X_{j-1}$ is an open manifold of dimension j. Set $U_k = X - X_{n-k}$ and let $i_k : U_k \hookrightarrow U_{k+1}$ denote the inclusion. Let $\bar{m}, \bar{n}$ be the lower and upper middle perversities, respectively.
Assume k is odd and $\mathbf{A}^\bullet \in D^b(U_k)$. Note that $\bar{n}(k) = \bar{m}(k) + 1$. Throughout this paper we will work with real coefficients.

Definition 2.1. The *lifting obstruction* $\mathcal{O}(\mathbf{A}^\bullet)$ associated with $\mathbf{A}^\bullet$ is defined to be

$$\mathcal{O}(\mathbf{A}^\bullet) = \mathbf{H}^{\bar{n}(k)-n}(Ri_{k*}\mathbf{A}^\bullet)[n - \bar{n}(k)] \in D^b(U_{k+1})$$

Thus, $\mathcal{O}(\mathbf{A}^\bullet)$ is a complex concentrated in dimension $\bar{n}(k) - n$. We shall henceforth use the shorthand notation

$$^{\bar{m}}\mathbf{A}^\bullet = \tau_{\leq \bar{m}(k)-n} Ri_{k*}\mathbf{A}^\bullet, {}^{\bar{n}}\mathbf{A}^\bullet = \tau_{\leq \bar{n}(k)-n} Ri_{k*}\mathbf{A}^\bullet, s = \bar{n}(k) - n.$$

Next, we construct a canonical distinguished triangle associated to $\mathcal{O}(\mathbf{A}^\bullet)$: By [**GM83**, prop. 1.15], there is a natural isomorphism

$$\begin{aligned}\Theta : \mathrm{Hom}_{D^b(U_{k+1})}({}^{\bar{n}}\mathbf{A}^\bullet, \mathcal{O}(\mathbf{A}^\bullet)) &\xrightarrow{\cong} \mathrm{Hom}(\mathbf{H}^s(Ri_{k*}\mathbf{A}^\bullet), \mathbf{H}^s(\mathcal{O}(\mathbf{A}^\bullet))) \\ &= \mathrm{Hom}(\mathbf{H}^s(Ri_{k*}\mathbf{A}^\bullet), \mathbf{H}^s(Ri_{k*}\mathbf{A}^\bullet))\end{aligned}$$

Define $\omega : {}^{\bar{n}}\mathbf{A}^\bullet \to \mathcal{O}(\mathbf{A}^\bullet)$ by

$$\omega = \Theta^{-1}(1_{\mathbf{H}^s(Ri_{k*}\mathbf{A}^\bullet)}).$$

Theorem 2.1. *There is a unique morphism* $\omega' : \mathcal{O}(\boldsymbol{A}^\bullet) \to {}^{\bar{m}}\boldsymbol{A}^\bullet[1]$ *such that the following is a distinguished triangle*

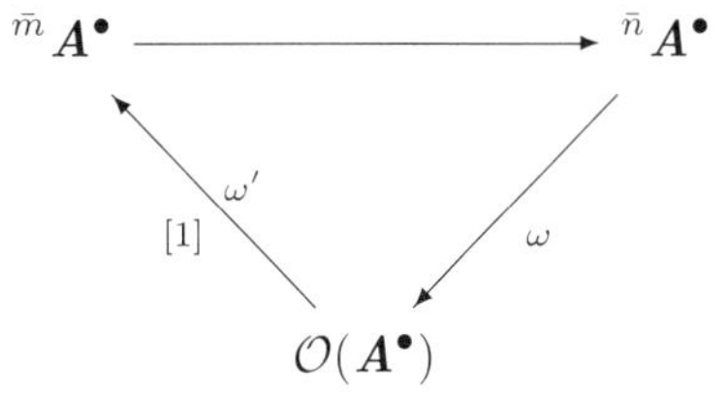

where ${}^{\bar{m}}\boldsymbol{A}^\bullet \to {}^{\bar{n}}\boldsymbol{A}^\bullet$ *is the canonical morphism.*

PROOF. Let

$$\begin{array}{ccc} {}^{\bar m}\mathbf{A}^\bullet & \longrightarrow & {}^{\bar n}\mathbf{A}^\bullet \\ & {\scriptstyle [1]}\nwarrow^{a_2} \quad \swarrow^{a_1} & \\ & \mathbf{M}^\bullet & \end{array}$$

be any distinguished triangle on the canonical morphism ${}^{\bar m}\mathbf{A}^\bullet \to {}^{\bar n}\mathbf{A}^\bullet$. As $\mathbf{H}^s({}^{\bar m}\mathbf{A}^\bullet) = \mathbf{H}^{s+1}({}^{\bar m}\mathbf{A}^\bullet) = 0$, the long exact sequence of cohomology sheaves induced by the above triangle implies that $\mathbf{H}^s(a_1)$ is an isomorphism. Using the isomorphism

$$\tilde{\Theta} : \mathrm{Hom}_{D^b(U_{k+1})}(\mathbf{M}^\bullet, \mathcal{O}(\mathbf{A}^\bullet)) \xrightarrow{\cong} \mathrm{Hom}(\mathbf{H}^s(\mathbf{M}^\bullet), \mathbf{H}^s(\mathcal{O}(\mathbf{A}^\bullet))),$$

set $\lambda = \tilde{\Theta}^{-1}(\mathbf{H}^s(\omega)\mathbf{H}^s(a_1)^{-1})$, $\lambda : \mathbf{M}^\bullet \to \mathcal{O}(\mathbf{A}^\bullet)$, so that

$$\mathbf{H}^s(\lambda a_1) = \mathbf{H}^s(\omega)\mathbf{H}^s(a_1)^{-1}\mathbf{H}^s(a_1) = \mathbf{H}^s(\omega)$$

and hence $\lambda a_1 = \omega$ by the injectivity of Θ. Now λ is a quasi-isomorphism as $\mathbf{H}^s(\lambda)$ is an isomorphism and $\mathbf{H}^i(\mathbf{M}^\bullet) = \mathbf{H}^i(\mathcal{O}(\mathbf{A}^\bullet)) = 0$ for $i \neq s$. It follows that with $\omega' = a_2\lambda^{-1}$ we have a commutative diagram

$$\begin{array}{ccccccc} {}^{\bar m}\mathbf{A}^\bullet & \longrightarrow & {}^{\bar n}\mathbf{A}^\bullet & \xrightarrow{a_1} & \mathbf{M}^\bullet & \xrightarrow{a_2} & {}^{\bar m}\mathbf{A}^\bullet[1] \\ \| & & \| & & \downarrow{\scriptstyle \lambda} & & \| \\ {}^{\bar m}\mathbf{A}^\bullet & \longrightarrow & {}^{\bar n}\mathbf{A}^\bullet & \xrightarrow{\omega} & \mathcal{O}(\mathbf{A}^\bullet) & \xrightarrow{\omega'} & {}^{\bar m}\mathbf{A}^\bullet[1] \end{array}$$

Since the upper row is a distinguished triangle and all vertical maps are isomorphisms, the lower row is distinguished as well.

As for the uniqueness of ω', observe that $\mathbf{H}^i({}^{\bar m}\mathbf{A}^\bullet[1]) = 0$ for $i > s - 2$ and thus

$$\mathrm{Hom}({}^{\bar m}\mathbf{A}^\bullet[1], \mathcal{O}(\mathbf{A}^\bullet)) = 0.$$

This implies the uniqueness of ω'. □

EXAMPLE 2.1. If, say, $\mathbf{A}^\bullet = \mathbf{IC}^\bullet_{\bar m}(U_3) = \mathbf{IC}^\bullet_{\bar n}(U_3)$ (here $\mathbf{IC}^\bullet_{\bar p}$ denotes the intersection homology complex for perversity $\bar p$), then $\tau_{\leq \bar m(3)-n} Ri_{3*}\mathbf{A}^\bullet = \mathbf{IC}^\bullet_{\bar m}(U_4)$ and $\tau_{\leq \bar n(3)-n} Ri_{3*}\mathbf{A}^\bullet = \mathbf{IC}^\bullet_{\bar n}(U_4)$. The lifting obstruction triangle becomes

$$\begin{array}{ccc} \mathbf{IC}^\bullet_{\bar m}(U_4) & \longrightarrow & \mathbf{IC}^\bullet_{\bar n}(U_4) \\ & {\scriptstyle [1]}\nwarrow \quad \swarrow & \\ & \mathcal{O}(\mathbf{A}^\bullet) & \end{array}$$

which on hypercohomology induces a long exact sequence

$$\ldots \to IH^{\bar m}_i(U_4) \to IH^{\bar n}_i(U_4) \to \mathcal{H}^{-i}(U_4; \mathcal{O}(\mathbf{A}^\bullet)) \to IH^{\bar m}_{i-1}(U_4) \to \ldots$$

So $\mathcal{H}^{-i}(U_4; \mathcal{O}(\mathbf{A}^\bullet))$ is the obstruction to lifting a class from $IH^{\bar n}_i(U_4)$ to $IH^{\bar m}_i(U_4)$, which explains our terminology.

REMARK 2.1. If $\mathbf{A}^\bullet$ satisfies a stalk condition $\mathbf{H}^i(\mathbf{A}^\bullet) = 0$ for $i > \bar{n}(k-1) - n$, then the support of the lifting obstruction $\mathcal{O}(\mathbf{A}^\bullet)$ is contained in $U_{k+1} - U_k = X_{n-k} - X_{n-k-1}$: Apply the functor i_k^* to the lifting obstruction triangle of theorem 2.1. As $i_k^* Ri_{k*} \cong 1$, we obtain

$$\begin{array}{ccc} \tau_{\leq \bar{m}(k)-n}\mathbf{A}^\bullet & \longrightarrow & \tau_{\leq \bar{n}(k)-n}\mathbf{A}^\bullet \\ & {\scriptstyle [1]}\nwarrow \quad \swarrow & \\ & i_k^*\mathcal{O}(\mathbf{A}^\bullet) & \end{array}$$

But $\bar{m}(k) = \bar{n}(k-1)$ and so $\tau_{\leq \bar{m}(k)-n}\mathbf{A}^\bullet \to \tau_{\leq \bar{n}(k)-n}\mathbf{A}^\bullet$ is a quasi-isomorphism, whence the restriction $\mathcal{O}(\mathbf{A}^\bullet)|_{U_k}$ is zero.

REMARK 2.2. One might attempt to *define* $\mathcal{O}(\mathbf{A}^\bullet)$ by a triangle on the canonical morphism ${}^{\bar{m}}\mathbf{A}^\bullet \to {}^{\bar{n}}\mathbf{A}^\bullet$. However, this involves an uncontrolled choice of morphisms ${}^{\bar{n}}\mathbf{A}^\bullet \to \mathcal{O}(\mathbf{A}^\bullet)$ and $\mathcal{O}(\mathbf{A}^\bullet) \to {}^{\bar{m}}\mathbf{A}^\bullet[1]$ (the latter is unique once the former is chosen) and the resulting triangle is not canonically associated with $\mathbf{A}^\bullet$.

The lifting obstruction is a covariant functor

$$\mathcal{O} : D^b(U_k) \longrightarrow D^b(U_{k+1}).$$

Given $f : \mathbf{A}^\bullet \to \mathbf{B}^\bullet$, let

$$\mathcal{O}(f) = \Theta^{-1}(\mathbf{H}^s(Ri_{k*}\mathbf{A}^\bullet) \xrightarrow{\mathbf{H}^s(Ri_{k*}f)} \mathbf{H}^s(Ri_{k*}\mathbf{B}^\bullet)),$$

where Θ is the isomorphism

$$\Theta : \mathrm{Hom}_{D^b(U_{k+1})}(\mathcal{O}(\mathbf{A}^\bullet), \mathcal{O}(\mathbf{B}^\bullet)) \xrightarrow{\cong} \mathrm{Hom}(\mathbf{H}^s(Ri_{k*}\mathbf{A}^\bullet), \mathbf{H}^s(Ri_{k*}\mathbf{B}^\bullet)).$$

Then $\mathcal{O}(1) = 1$ and $\mathcal{O}(gf) = \mathcal{O}(g)\mathcal{O}(f)$ for $g : \mathbf{B}^\bullet \to \mathbf{C}^\bullet$ are immediate.
In fact, one can say more:

PROPOSITION 2.1. *The following is a commutative diagram:*

$$\begin{array}{ccc} {}^{\bar{n}}\boldsymbol{A}^\bullet & \xrightarrow{\ \omega\ } & \mathcal{O}(\boldsymbol{A}^\bullet) \\ {\scriptstyle {}^{\bar{n}}f}\downarrow & & \downarrow{\scriptstyle \mathcal{O}(f)} \\ {}^{\bar{n}}\boldsymbol{B}^\bullet & \xrightarrow{\ \omega\ } & \mathcal{O}(\boldsymbol{B}^\bullet) \end{array}$$

PROOF. Simply note that $\mathbf{H}^s(\mathcal{O}(f)\omega) = \mathbf{H}^s(\omega^{\bar{n}}f)$ and use the isomorphism

$$\mathrm{Hom}_{D^b(U_{k+1})}({}^{\bar{n}}\mathbf{A}^\bullet, \mathcal{O}(\mathbf{B}^\bullet)) \cong \mathrm{Hom}(\mathbf{H}^s({}^{\bar{n}}\mathbf{A}^\bullet), \mathbf{H}^s(Ri_{k*}\mathbf{B}^\bullet)).$$

□

2. The Category of Self-Dual Sheaves Compatible with IH

Assume X is oriented, i.e. assume we can and do fix an isomorphism $\mathbb{D}^\bullet_{U_2} \xrightarrow{\cong} \mathbb{R}_{U_2}[n]$. We assume furthermore that the strata and links are also given orientations that are compatible with the orientation of X under the stratum-preserving homeomorphism $U \times c^\circ L_x \xrightarrow{\cong} N_x$, where $x \in X_{n-k} - X_{n-k-1}$, U is an open neighborhood of x in $X_{n-k} - X_{n-k-1}$, $c^\circ L_x$ is the open cone on the link L_x at x and N_x is a neighborhood of x in X. We define the category of complexes of sheaves suitable for studying intersection-homology type invariants on non-Witt spaces. The objects of this category should satisfy two properties: On the one hand, they should be self-dual, on the other hand, they should be as close to the middle perversity intersection chain sheaves as possible, that is, interpolate between $\mathbf{IC}^\bullet_{\bar{m}}(X)$ and $\mathbf{IC}^\bullet_{\bar{n}}(X)$. Given these specifications, we adopt the following definition:

DEFINITION 2.2. Let $SD(X)$ be the full subcategory of $D^b(X)$ whose objects $\mathbf{L}^\bullet$ satisfy the following axioms:

(SD1): Normalization: $\mathbf{L}^\bullet$ has an associated isomorphism $\nu : \mathbb{R}_{U_2}[n] \xrightarrow{\cong} \mathbf{L}^\bullet|_{U_2}$.
(SD2): Lower bound: $\mathbf{H}^i(\mathbf{L}^\bullet) = 0$, for $i < -n$.
(SD3): Stalk condition for the upper middle perversity $\bar{n}$: $\mathbf{H}^i(\mathbf{L}^\bullet|_{U_{k+1}}) = 0$, for $i > \bar{n}(k) - n, k \geq 2$.
(SD4): Self-Duality: $\mathbf{L}^\bullet$ has an associated isomorphism $d : \mathcal{D}\mathbf{L}^\bullet[n] \xrightarrow{\cong} \mathbf{L}^\bullet$ such that $\mathcal{D}d[n] = d$ and $d|_{U_2}$ is compatible to the orientation under normalization so that

$$\begin{array}{ccc} \mathbb{R}_{U_2}[n] & \xrightarrow[\cong]{\nu} & \mathbf{L}^\bullet|_{U_2} \\ \text{orient} \uparrow \cong & & \cong \uparrow d|_{U_2} \\ \mathbb{D}^\bullet_{U_2} & \xrightarrow[\cong]{\mathcal{D}\nu^{-1}[n]} & \mathcal{D}\mathbf{L}^\bullet|_{U_2}[n] \end{array}$$

commutes.

The reader may want to compare this set of axioms with [AX1] and [AX3] of [**GM83**]. Note that if X is a Witt space and $\mathbf{L}^\bullet \in SD(X)$, then $\mathbf{IC}^\bullet_{\bar{m}}(X) \cong \mathbf{L}^\bullet \cong \mathbf{IC}^\bullet_{\bar{n}}(X)$. We list some facts about $SD(X)$:

LEMMA 2.1. *If $\mathbf{L}^\bullet \in SD(X)$, then $\mathbf{L}^\bullet$ satisfies the $\bar{m}$-costalk condition, i.e.*

$$\mathbf{H}^i(j_x^! \mathbf{L}^\bullet) = 0 \quad \textit{for } i \leq \bar{m}(k) - k + 1, x \in X_{n-k} - X_{n-k-1},$$

where $j_x : \{x\} \hookrightarrow X$ is the inclusion of a point.

PROOF. Using (**SD4**),

$$\mathbf{H}^i(j_x^! \mathbf{L}^\bullet) \cong \mathbf{H}^i(j_x^! \mathcal{D}\mathbf{L}^\bullet[n]) \cong \mathbf{H}^{i+n}(\mathcal{D}(j_x^* \mathbf{L}^\bullet)) \cong$$

$$\mathrm{Hom}(\mathbf{H}^{-i-n}(\mathbf{L}^\bullet)_x, \mathbb{R}) = 0$$

for $-i - n > \bar{n}(k) - n$, by (**SD3**). This condition is equivalent to $i < -\bar{n}(k) = \bar{m}(k) - k + 2$.

□

LEMMA 2.2. *If $\mathbf{E}^\bullet, \mathbf{F}^\bullet \in SD(U_{k+1})$, then restriction induces an injection*

$$\mathrm{Hom}_{D^b(U_{k+1})}(\mathbf{E}^\bullet, \mathbf{F}^\bullet) \hookrightarrow \mathrm{Hom}_{D^b(U_k)}(\mathbf{E}^\bullet|_{U_k}, \mathbf{F}^\bullet|_{U_k}).$$

PROOF. By (**SD3**), $\mathbf{H}^i(\mathbf{E}^\bullet|_{U_{k+1}}) = 0$, $i > \bar{n}(k) - n$. By lemma 2.1, $\mathbf{H}^i(j_x^! \mathbf{F}^\bullet) = 0$, $i \leq \bar{n}(k) - k$. The statement follows from (1.3) of [**CS91**].

□

LEMMA 2.3. *If* $\mathbf{A}^\bullet \in SD(U_k)$, *then* $d : \mathcal{D}\mathbf{A}^\bullet[n] \cong \mathbf{A}^\bullet$ *induces a self-duality isomorphism* δ *for the lifting obstruction:*

$$\delta : \mathcal{D}\mathcal{O}(\mathbf{A}^\bullet)[n+1] \xrightarrow{\cong} \mathcal{O}(\mathbf{A}^\bullet).$$

PROOF. Start out with the canonical triangle given by theorem 2.1:

$$^{\bar{m}}\mathbf{A}^\bullet \xrightarrow{c} {}^{\bar{n}}\mathbf{A}^\bullet \xrightarrow{\omega} \mathcal{O}(\mathbf{A}^\bullet) \xrightarrow{\omega'} {}^{\bar{m}}\mathbf{A}^\bullet[1],$$

where c denotes the canonical morphism. Dualizing and shifting by $[n]$ yields the triangle

$$\mathcal{D}^{\bar{n}}\mathbf{A}^\bullet[n] \xrightarrow{\mathcal{D}c[n]} \mathcal{D}^{\bar{m}}\mathbf{A}^\bullet[n] \xrightarrow{\mathcal{D}\omega'[n+1]} \mathcal{D}\mathcal{O}(\mathbf{A}^\bullet)[n+1] \xrightarrow{(-1)^{n+1}\mathcal{D}\omega[n+1]}$$

By section 9 of [**GM83**], d induces isomorphisms $d' : \mathcal{D}^{\bar{m}}\mathbf{A}^\bullet[n] \xrightarrow{\cong} {}^{\bar{n}}\mathbf{A}^\bullet$ and $d'' : \mathcal{D}^{\bar{n}}\mathbf{A}^\bullet[n] \xrightarrow{\cong} {}^{\bar{m}}\mathbf{A}^\bullet$ $(\mathcal{D}d''[n] = d')$ and we have a commutative diagram

$$\begin{array}{ccc}
\mathcal{D}^{\bar{n}}\mathbf{A}^\bullet[n] & \xrightarrow{\mathcal{D}c[n]} & \mathcal{D}^{\bar{m}}\mathbf{A}^\bullet[n] \\
{\scriptstyle d''}\downarrow{\scriptstyle \cong} & & {\scriptstyle d'}\downarrow{\scriptstyle \cong} \\
{}^{\bar{m}}\mathbf{A}^\bullet & \xrightarrow{c} & {}^{\bar{n}}\mathbf{A}^\bullet
\end{array}$$

which can be embedded in an isomorphism of triangles

$$\begin{array}{ccccccc}
\mathcal{D}^{\bar{n}}\mathbf{A}^\bullet[n] & \xrightarrow{\mathcal{D}c[n]} & \mathcal{D}^{\bar{m}}\mathbf{A}^\bullet[n] & \xrightarrow{\mathcal{D}\omega'[n+1]} & \mathcal{D}\mathcal{O}(\mathbf{A}^\bullet)[n+1] & \xrightarrow{(-1)^{n+1}\mathcal{D}\omega[n+1]} & \\
\downarrow{\scriptstyle \cong} & & {\scriptstyle d'}\downarrow{\scriptstyle \cong} & & {\scriptstyle \delta}\downarrow{\scriptstyle \cong} & & \\
{}^{\bar{m}}\mathbf{A}^\bullet & \xrightarrow{c} & {}^{\bar{n}}\mathbf{A}^\bullet & \xrightarrow{\omega} & \mathcal{O}(\mathbf{A}^\bullet) & \xrightarrow{\omega'} &
\end{array}$$

□

LEMMA 2.4. $\delta : \mathcal{D}\mathcal{O}(\mathbf{A}^\bullet)[n+1] \xrightarrow{\cong} \mathcal{O}(\mathbf{A}^\bullet)$ *constructed in the previous lemma is the unique morphism such that*

$$\begin{array}{ccc}
\mathcal{D}^{\bar{m}}\mathbf{A}^\bullet[n] & \xrightarrow{\mathcal{D}\omega'[n+1]} & \mathcal{D}\mathcal{O}(\mathbf{A}^\bullet)[n+1] \\
{\scriptstyle d'}\downarrow{\scriptstyle \cong} & & {\scriptstyle \delta}\downarrow{\scriptstyle \cong} \\
{}^{\bar{n}}\mathbf{A}^\bullet & \xrightarrow{\omega} & \mathcal{O}(\mathbf{A}^\bullet)
\end{array}$$

commutes.

PROOF. As

$$\mathrm{Hom}_{D^b(U_{k+1})}(\mathcal{D}\mathcal{O}(\mathbf{A}^\bullet)[n+1], \mathcal{O}(\mathbf{A}^\bullet)) \xrightarrow{\cong} \mathrm{Hom}(\mathbf{H}^s(\mathcal{D}\mathcal{O}(\mathbf{A}^\bullet)[n+1]), \mathbf{H}^s(\mathcal{O}(\mathbf{A}^\bullet))),$$

it suffices to show that $\mathbf{H}^s(\delta)$ is the unique map which makes the induced square of cohomology sheaves in dimension s commute. Now $\mathbf{H}^s(\omega)$ is an isomorphism.

Hence, by commutativity, $\mathbf{H}^s(\mathcal{D}\omega'[n+1])$ is an isomorphism and

$$\mathbf{H}^s(\delta) = \mathbf{H}^s(\omega)\mathbf{H}^s(d')\mathbf{H}^s(\mathcal{D}\omega'[n+1])^{-1}.$$

□

The above two lemmas tell us that to every $(\mathbf{A}^\bullet, d) \in SD(U_k)$ we can uniquely associate an isomorphism $\delta = \delta(d) : \mathcal{D}\mathcal{O}(\mathbf{A}^\bullet)[n+1] \xrightarrow{\cong} \mathcal{O}(\mathbf{A}^\bullet)$.

Given $\mathbf{L}^\bullet \in SD(U_{k+1})$, we shall describe certain natural morphisms

$${}^{\bar{m}}(\mathbf{L}^\bullet|_{U_k}) \xrightarrow{a} \mathbf{L}^\bullet \xrightarrow{b} {}^{\bar{n}}(\mathbf{L}^\bullet|_{U_k}).$$

Consider the adjunction $\mathbf{L}^\bullet \to Ri_{k*}i_k^*\mathbf{L}^\bullet$. Over U_k, this is an isomorphism. For the inclusion $j_k : U_{k+1} - U_k \hookrightarrow U_{k+1}$, we have that

$$\mathbf{H}^i(j_k^*\mathbf{L}^\bullet) \to \mathbf{H}^i(j_k^*Ri_{k*}i_k^*\mathbf{L}^\bullet)$$

is an isomorphism when $i \leq \bar{m}(k) - n$, as by lemma 2.1, $\mathbf{L}^\bullet$ satisfies the $\bar{m}$-costalk condition. It follows that

$$\tau_{\leq \bar{m}(k)-n}\mathbf{L}^\bullet \xrightarrow{\cong} \tau_{\leq \bar{m}(k)-n}Ri_{k*}i_k^*\mathbf{L}^\bullet$$

is an isomorphism.

We define a to be the composition of the inverse morphism ${}^{\bar{m}}\mathbf{L}^\bullet|_{U_k} \xrightarrow{\cong} \tau_{\leq \bar{m}(k)-n}\mathbf{L}^\bullet$ with the canonical map $\tau_{\leq \bar{m}(k)-n}\mathbf{L}^\bullet \to \mathbf{L}^\bullet$:

$$\begin{array}{ccc} {}^{\bar{m}}\mathbf{L}^\bullet|_{U_k} & \xrightarrow{\cong} & \tau_{\leq \bar{m}(k)-n}\mathbf{L}^\bullet \\ & \searrow^{a} & \downarrow \\ & & \mathbf{L}^\bullet \end{array}$$

Note that on U_k, a is the identity: $a|_{U_k} = 1_{\mathbf{L}^\bullet|_{U_k}}$.

To construct b, consider the canonical map $\tau_{\leq \bar{n}(k)-n}\mathbf{L}^\bullet \to \mathbf{L}^\bullet$. This is a quasi-isomorphism by (**SD3**). Adjunction induces $\tau_{\leq \bar{n}(k)-n}\mathbf{L}^\bullet \to \tau_{\leq \bar{n}(k)-n}Ri_{k*}i_k^*\mathbf{L}^\bullet$ and we define b to be the composition

$$\begin{array}{ccc} \mathbf{L}^\bullet & \xrightarrow{\cong} & \tau_{\leq \bar{n}(k)-n}\mathbf{L}^\bullet \\ & \searrow^{b} & \downarrow \\ & & {}^{\bar{n}}\mathbf{L}^\bullet|_{U_k} \end{array}$$

Note that on U_k, b is the identity: $b|_{U_k} = 1_{\mathbf{L}^\bullet|_{U_k}}$.

The following result clarifies the relation between intersection homology sheaves and objects of $SD(X)$.

THEOREM 2.2. *Given* $\boldsymbol{L}^\bullet \in SD(X)$, *there exist morphisms*

$$\boldsymbol{IC}^\bullet_{\bar{m}}(X) \xrightarrow{\alpha} \boldsymbol{L}^\bullet \xrightarrow{\beta} \boldsymbol{IC}^\bullet_{\bar{n}}(X)$$

such that

1. *$\alpha|_{U_2} = \nu : \mathbb{R}_{U_2}[n] \xrightarrow{\cong} \boldsymbol{L}^\bullet|_{U_2}$ (where ν is the normalization)*
   $\beta|_{U_2} = \nu^{-1} : \boldsymbol{L}^\bullet|_{U_2} \xrightarrow{\cong} \mathbb{R}_{U_2}[n]$
2. *α and β are unique with respect to property 1.,*
3. *the following diagram is commutative*

$$\begin{array}{ccc} \boldsymbol{IC}^\bullet_{\bar m}(X) & \xrightarrow{\alpha} & \boldsymbol{L}^\bullet \\ \uparrow{\scriptstyle\cong} & & \uparrow{\scriptstyle\cong\, d} \\ \mathcal{D}\boldsymbol{IC}^\bullet_{\bar n}(X)[n] & \xrightarrow{\mathcal{D}\beta[n]} & \mathcal{D}\boldsymbol{L}^\bullet[n] \end{array}$$

   *(where d is given by (**SD4**).)*
4. *α, β are functorial with respect to morphisms whose restriction to U_2 is compatible with the normalizations. More precisely, given $f : \boldsymbol{L}^\bullet_1 \to \boldsymbol{L}^\bullet_2$, $\boldsymbol{L}^\bullet_1, \boldsymbol{L}^\bullet_2 \in SD(X)$, such that*

$$\begin{array}{ccc} & & \boldsymbol{L}^\bullet_1|_{U_2} \\ & \overset{\nu_1}{\nearrow} & \\ \mathbb{R}_{U_2}[n] & & \downarrow{\scriptstyle f|_{U_2}} \\ & \underset{\nu_2}{\searrow} & \\ & & \boldsymbol{L}^\bullet_2|_{U_2} \end{array}$$

   commutes, there is a commutative diagram

$$\begin{array}{ccccc} & & \boldsymbol{L}^\bullet_1 & & \\ & \overset{\alpha_1}{\nearrow} & & \overset{\beta_1}{\searrow} & \\ \boldsymbol{IC}^\bullet_{\bar m}(X) & & \downarrow{\scriptstyle f} & & \boldsymbol{IC}^\bullet_{\bar n}(X) \\ & \underset{\alpha_2}{\searrow} & & \underset{\beta_2}{\nearrow} & \\ & & \boldsymbol{L}^\bullet_2 & & \end{array}$$

PROOF. We will construct α and β inductively. On U_2, $\mathcal{D}\mathbf{IC}^\bullet_{\bar n}(U_2)[n] \xrightarrow{\cong} \mathbf{IC}^\bullet_{\bar m}(U_2)$ is the orientation $\mathbb{D}^\bullet_{U_2} \to \mathbb{R}_{U_2}[n]$. Hence by (**SD4**), we have a commutative diagram

$$\begin{array}{ccc} \mathbf{IC}^\bullet_{\bar m}(U_2) & \xrightarrow{\nu} & \mathbf{L}^\bullet|_{U_2} \\ \uparrow{\scriptstyle\phi_2} & & \uparrow{\scriptstyle d|_{U_2}} \\ \mathcal{D}\mathbf{IC}^\bullet_{\bar n}(U_2)[n] & \xrightarrow{\mathcal{D}\nu^{-1}[n]} & \mathcal{D}\mathbf{L}^\bullet|_{U_2}[n] \end{array}$$

(where $\phi_2 =$ orientation.)
Therefore, we set $\alpha_2 = \nu$ and $\beta_2 = \nu^{-1}$ on U_2. Now assume inductively that the

commutative diagram

$$\begin{array}{ccc}
\mathbf{IC}^{\bullet}_{\bar{m}}(U_k) & \xrightarrow{\alpha_k} & \mathbf{L}^{\bullet}|_{U_k} \\
\phi_k \Big\uparrow & & \Big\uparrow d|_{U_k} \\
\mathcal{D}\mathbf{IC}^{\bullet}_{\bar{n}}(U_k)[n] & \xrightarrow{\mathcal{D}\beta_k[n]} & \mathcal{D}\mathbf{L}^{\bullet}|_{U_k}[n]
\end{array}$$

has been constructed. By [**GM83**], section 9, there exists a unique isomorphism

$$\phi_{k+1} : \mathcal{D}(\tau_{\leq \bar{n}(k)-n} Ri_{k*}\mathbf{IC}^{\bullet}_{\bar{n}}(U_k))[n] \xrightarrow{\cong} \tau_{\leq \bar{m}(k)-n} Ri_{k*}\mathbf{IC}^{\bullet}_{\bar{m}}(U_k)$$

such that $\phi_{k+1}|_{U_k} = \phi_k$. This yields $\phi_{k+1} : \mathcal{D}\mathbf{IC}^{\bullet}_{\bar{n}}(U_{k+1})[n] \xrightarrow{\cong} \mathbf{IC}^{\bullet}_{\bar{m}}(U_{k+1})$. Similarly, there exists a unique isomorphism

$$d' : \mathcal{D}(\tau_{\leq \bar{n}(k)-n} Ri_{k*}\mathbf{L}^{\bullet}|_{U_k})[n] \xrightarrow{\cong} \tau_{\leq \bar{m}(k)-n} Ri_{k*}\mathbf{L}^{\bullet}|_{U_k}$$

such that $d'|_{U_k} = d|_{U_k}$. We verify that the following diagram commutes (we shall write $\tau_{\leq \bar{m}}$ for $\tau_{\leq \bar{m}(k)-n}$ and $\tau_{\leq \bar{n}}$ for $\tau_{\leq \bar{n}(k)-n}$):

$$\begin{array}{ccc}
\mathbf{IC}^{\bullet}_{\bar{m}}(U_{k+1}) & \xrightarrow{\tau_{\leq \bar{m}} Ri_{k*}\alpha_k} & {}^{\bar{m}}\mathbf{L}^{\bullet}|_{U_k} \\
\phi_{k+1} \Big\uparrow \cong & & \cong \Big\uparrow d' \\
\mathcal{D}\mathbf{IC}^{\bullet}_{\bar{n}}(U_{k+1})[n] & \xrightarrow{\mathcal{D}(\tau_{\leq \bar{n}} Ri_{k*}\beta_k)[n]} & \mathcal{D}({}^{\bar{n}}\mathbf{L}^{\bullet}|_{U_k})[n]
\end{array}$$

Let $\mathbf{X}^{\bullet} = \mathcal{D}\mathbf{IC}^{\bullet}_{\bar{n}}(U_{k+1})[n]$, $\mathbf{Y}^{\bullet} = {}^{\bar{m}}\mathbf{L}^{\bullet}|_{U_k}$. We have $\mathbf{H}^i(\mathbf{X}^{\bullet}_{U_{l+1}}) = 0$ for $i > \bar{m}(l) - n$, $2 \leq l \leq k$ and $\mathbf{H}^i(j_x^! \mathbf{Y}^{\bullet}) = 0$ for $i \leq \bar{m}(l) - l + 1$, $x \in X_{n-l} - X_{n-l-1}$, $2 \leq l \leq k$. By [**CS91**] (1.3), restriction induces an isomorphism

$$\operatorname{Hom}_{D^b(U_{k+1})}(\mathbf{X}^{\bullet}, \mathbf{Y}^{\bullet}) \xrightarrow{\cong} \operatorname{Hom}_{D^b(U_k)}(\mathbf{X}^{\bullet}|_{U_k}, \mathbf{Y}^{\bullet}|_{U_k}),$$

that is,

$$\operatorname{Hom}_{D^b(U_{k+1})}(\mathcal{D}\mathbf{IC}^{\bullet}_{\bar{n}}(U_{k+1})[n], {}^{\bar{m}}\mathbf{L}^{\bullet}|_{U_k}) \xrightarrow{\cong} \operatorname{Hom}_{D^b(U_k)}(\mathcal{D}\mathbf{IC}^{\bullet}_{\bar{n}}(U_k)[n], \mathbf{L}^{\bullet}|_{U_k})$$

(using $i_k^* {}^{\bar{m}}\mathbf{L}^{\bullet}|_{U_k} = i_k^* \tau_{\leq \bar{m}} Ri_{k*}\mathbf{L}^{\bullet}|_{U_k} \cong \tau_{\leq \bar{m}} i_k^* Ri_{k*}\mathbf{L}^{\bullet}|_{U_k} \cong \tau_{\leq \bar{m}(k)-n}\mathbf{L}^{\bullet}|_{U_k} \cong \tau_{\leq \bar{n}(k-1)-n}\mathbf{L}^{\bullet}|_{U_k} \cong \mathbf{L}^{\bullet}|_{U_k}$ by (**SD3**).)
Observe that $(\tau_{\leq \bar{m}} Ri_{k*}\alpha_k)|_{U_k} = \alpha_k$ and as $i_k^*\mathcal{D} = \mathcal{D} i_k^! = \mathcal{D} i_k^*$, $i_k^*\mathcal{D}(\tau_{\leq \bar{n}} Ri_{k*}\beta_k)[n] = \mathcal{D}(i_k^* \tau_{\leq \bar{n}} Ri_{k*}\beta_k)[n] = \mathcal{D}(\tau_{\leq \bar{n}}\beta_k)[n] = \mathcal{D}\beta_k[n]$.
We calculate

$$\begin{aligned}
i_k^*(\tau_{\leq \bar{m}} Ri_{k*}\alpha_k \circ \phi_{k+1}) &= (\tau_{\leq \bar{m}} Ri_{k*}\alpha_k)|_{U_k} \circ (\phi_{k+1}|_{U_k}) \\
&= \alpha_k \phi_k \\
&= (d|_{U_k}) \circ \mathcal{D}\beta_k[n] \\
&= (d'|_{U_k}) \circ \mathcal{D}(\tau_{\leq \bar{n}} Ri_{k*}\beta_k)[n]|_{U_k} \\
&= i_k^*(d' \circ \mathcal{D}(\tau_{\leq \bar{n}} Ri_{k*}\beta_k)[n]).
\end{aligned}$$

By the injectivity of i_k^*,

$$\tau_{\leq \bar{m}} Ri_{k*}\alpha_k \circ \phi_{k+1} = d' \circ \mathcal{D}(\tau_{\leq \bar{n}} Ri_{k*}\beta_k)[n]$$

which completes the verification of commutativity.

Next, we claim that the diagram below is commutative as well:

$$\begin{array}{ccc}
{}^{\bar{m}}\mathbf{L}^\bullet|_{U_k} & \xrightarrow{\;a\;} & \mathbf{L}^\bullet|_{U_{k+1}} \\
\uparrow {\scriptstyle d'}\,{\scriptstyle\cong} & & \uparrow {\scriptstyle d|_{U_{k+1}}} \\
\mathcal{D}({}^{\bar{n}}\mathbf{L}^\bullet|_{U_k})[n] & \xrightarrow{\mathcal{D}b[n]} & \mathcal{D}\mathbf{L}^\bullet|_{U_{k+1}}[n]
\end{array}$$

(where a, b are the maps defined prior to the statement of the theorem.)

To verify the claim, we use the same method as above: Let $\mathbf{X}^\bullet = \mathcal{D}({}^{\bar{n}}\mathbf{L}^\bullet|_{U_k})[n]$, $\mathbf{Y}^\bullet = \mathbf{L}^\bullet|_{U_{k+1}}$.

Then $\mathbf{H}^i(\mathbf{X}^\bullet_{U_{l+1}}) = 0$ for $i > \bar{n}(l) - n$, $2 \leq l \leq k$ and $\mathbf{H}^i(j_x^!\mathbf{Y}^\bullet) = 0$ for $i \leq \bar{n}(l) - l$, $x \in X_{n-l} - X_{n-l-1}$, $2 \leq l \leq k$. Again invoking [**CS91**] (1.3), restriction induces an injection

$$\mathrm{Hom}_{D^b(U_{k+1})}(\mathbf{X}^\bullet, \mathbf{Y}^\bullet) \overset{i_k^*}{\hookrightarrow} \mathrm{Hom}_{D^b(U_k)}(\mathbf{X}^\bullet|_{U_k}, \mathbf{Y}^\bullet|_{U_k}).$$

We have $i_k^*\mathcal{D}b[n] = \mathcal{D}(b|_{U_k})[n] = \mathcal{D}(1_{\mathbf{L}^\bullet|U_k})[n] = 1_{\mathcal{D}\mathbf{L}^\bullet|U_k[n]}$, and so we compute

$$\begin{aligned}
i_k^*(ad') &= a|_{U_k} \circ d'|_{U_k} = 1_{\mathbf{L}^\bullet|U_k} \circ d|_{U_k} = d|_{U_k} \\
&= d|_{U_k} \circ 1_{\mathcal{D}\mathbf{L}^\bullet|U_k[n]} = d|_{U_k} \circ i_k^*\mathcal{D}b[n] \\
&= i_k^*(d|_{U_{k+1}}\mathcal{D}b[n]).
\end{aligned}$$

By injectivity of i_k^*,

$$ad' = d|_{U_{k+1}}\mathcal{D}b[n]$$

and the claim is verified. To obtain $\alpha_{k+1}, \beta_{k+1}$, we put the two squares together:

$$\begin{array}{ccccc}
\mathbf{IC}^\bullet_{\bar{m}}(U_{k+1}) & \xrightarrow{\tau_{\leq\bar{m}}Ri_{k*}\alpha_k} & {}^{\bar{m}}\mathbf{L}^\bullet|_{U_k} & \xrightarrow{\;a\;} & \mathbf{L}^\bullet|_{U_{k+1}} \\
\uparrow {\scriptstyle \phi_{k+1}}\,{\scriptstyle\cong} & & \uparrow {\scriptstyle\cong}\,{\scriptstyle d'} & & \uparrow {\scriptstyle\cong}\,{\scriptstyle d|_{U_{k+1}}} \\
\mathcal{D}\mathbf{IC}^\bullet_{\bar{n}}(U_{k+1})[n] & \xrightarrow{\mathcal{D}(\tau_{\leq\bar{n}}Ri_{k*}\beta_k)[n]} & \mathcal{D}({}^{\bar{n}}\mathbf{L}^\bullet|_{U_k})[n] & \xrightarrow{\mathcal{D}b[n]} & \mathcal{D}\mathbf{L}^\bullet|_{U_{k+1}}[n]
\end{array}$$

Define

$$\begin{aligned}
\alpha_{k+1} &= a \circ \tau_{\leq\bar{m}}Ri_{k*}\alpha_k, \\
\beta_{k+1} &= (\tau_{\leq\bar{n}}Ri_{k*}\beta_k) \circ b.
\end{aligned}$$

This concludes the construction of α, β such that 1. and 3. hold.

To see 2., observe $\mathbf{H}^i(\mathbf{IC}^\bullet_{\bar{m}}(X)|_{U_{k+1}}) = 0$ for $i > \bar{m}(k) - n, k \geq 2$, and $\mathbf{H}^i(j_x^!\mathbf{L}^\bullet) = 0$ for $i \leq \bar{m}(k) - k + 1$, $x \in X_{n-k} - X_{n-k-1}$. By [**CS91**] (1.3), restriction to U_2 induces an isomorphism

$$\mathrm{Hom}_{D^b(X)}(\mathbf{IC}^\bullet_{\bar{m}}(X), \mathbf{L}^\bullet) \xrightarrow{\cong} \mathrm{Hom}_{D^b(U_2)}(\mathbb{R}_{U_2}[n], \mathbf{L}^\bullet|_{U_2}).$$

Hence α is unique. Similarly, $\mathbf{H}^i(\mathbf{L}^\bullet|_{U_{k+1}}) = 0$ for $i > \bar{n}(k) - n, k \geq 2$, and $\mathbf{H}^i(j_x^!\mathbf{IC}^\bullet_{\bar{n}}(X)) = 0$ for $i \leq \bar{n}(k) - k + 1$, $x \in X_{n-k} - X_{n-k-1}$. Again restriction to U_2 induces an isomorphism

$$\mathrm{Hom}_{D^b(X)}(\mathbf{L}^\bullet, \mathbf{IC}^\bullet_{\bar{n}}(X)) \xrightarrow{\cong} \mathrm{Hom}_{D^b(U_2)}(\mathbf{L}^\bullet|_{U_2}, \mathbb{R}_{U_2}[n]),$$

so β is unique.
The functoriality of α and β follows readily from uniqueness: $(f\alpha_1)|_{U_2} = \alpha_2|_{U_2}$ by assumption and using 1. By uniqueness, $f\alpha_1 = \alpha_2$. Similarly, $\beta_2 f = \beta_1$.

□

3. Lagrangian Structures

Theorem 2.2 shows that the objects of $SD(X)$ are the closest self-dual "approximation" to intersection homology possible on non-Witt spaces. We are thus interested in understanding the structure of sheaves in $SD(X)$. The main tool in this analysis and indeed the building blocks for objects in $SD(X)$ are certain locally constant sheaves that we call Lagrangian structures and which we will regard as complexes in the derived category concentrated in one dimension.
Fix some k odd, $k \geq 2$. We state the definition first and then give a detailed discussion.

Definition 2.3. Let $\mathbf{A}^\bullet \in SD(U_k)$. Recall that the lifting obstruction $\mathcal{O}(\mathbf{A}^\bullet)$ is self-dual by lemma 2.3.
A *Lagrangian structure* is a morphism $\mathcal{L} \longrightarrow \mathcal{O}(\mathbf{A}^\bullet)$, $\mathcal{L} \in D^b(U_{k+1})$, which induces injections on stalks and has the property that some distinguished triangle on $\mathcal{L} \longrightarrow \mathcal{O}(\mathbf{A}^\bullet)$ is a nullcobordism (in the sense of [**CS91**]) for the self-dual lifting obstruction.

This means precisely the following: First of all, some triangle on $\phi : \mathcal{L} \longrightarrow \mathcal{O}(\mathbf{A}^\bullet)$ has to be of the form

$$\begin{array}{ccc} \mathcal{L} & \xrightarrow{\quad\phi\quad} & \mathcal{O}(\mathbf{A}^\bullet) \\ & {\scriptstyle [1]}\ \nwarrow^{\gamma} \qquad \swarrow_{\phi'} & \\ & \mathcal{D}\mathcal{L}[n+1] & \end{array} \tag{3}$$

and we require $\mathcal{D}\gamma[n+1] = \gamma[-1]$.
Now γ induces a class $\{d_\gamma\}$ of isomorphisms $d_\gamma : \mathcal{D}\mathcal{O}(\mathbf{A}^\bullet)[n+1] \xrightarrow{\cong} \mathcal{O}(\mathbf{A}^\bullet)$ as follows:
After dualizing and shifting the triangle

$$\mathcal{D}\mathcal{L}[n] \xrightarrow{-\gamma[-1]} \mathcal{L} \xrightarrow{\phi} \mathcal{O}(\mathbf{A}^\bullet) \xrightarrow{\phi'} \mathcal{D}\mathcal{L}[n+1]$$

we obtain

$$\mathcal{D}\mathcal{L}[n] \xrightarrow{-\mathcal{D}\gamma[n+1]} \mathcal{L} \xrightarrow{\mathcal{D}\phi'[n+1]} \mathcal{D}\mathcal{O}(\mathbf{A}^\bullet)[n+1] \xrightarrow{(-1)^{n+1}\mathcal{D}\phi[n+1]} \mathcal{D}\mathcal{L}[n+1].$$

The self-duality of γ may be represented as a commutative square

$$\begin{array}{ccc} \mathcal{D}\mathcal{L}[n] & \xrightarrow{-\mathcal{D}\gamma[n+1]} & \mathcal{L} \\ \Vert & & \Vert \\ \mathcal{D}\mathcal{L}[n] & \xrightarrow{-\gamma[-1]} & \mathcal{L} \end{array}$$

which upon embedding in an isomorphism of triangles

$$
(4)\qquad
\begin{array}{ccccccc}
\mathcal{DL}[n] & \xrightarrow{-\mathcal{D}\gamma[n+1]} & \mathcal{L} & \xrightarrow{\mathcal{D}\phi'[n+1]} & \mathcal{DO}(\mathbf{A}^\bullet)[n+1] & \xrightarrow{(-1)^{n+1}\mathcal{D}\phi[n+1]} & \mathcal{DL}[n+1] \\
\| & & \| & & \cong \downarrow d_\gamma & & \| \\
\mathcal{DL}[n] & \xrightarrow{-\gamma[-1]} & \mathcal{L} & \xrightarrow{\phi} & \mathcal{O}(\mathbf{A}^\bullet) & \xrightarrow{\phi'} & \mathcal{DL}[n+1]
\end{array}
$$

induces (non-uniquely) some isomorphism d_γ.
In order for the triangle (3) to be a nullcobordism for $\mathcal{O}(\mathbf{A}^\bullet)$, one requires that $\delta \in \{d_\gamma\}$, where δ is the isomorphism from lemma 2.3, induced by $d : \mathcal{D}\mathbf{A}^\bullet[n] \cong \mathbf{A}^\bullet$.

DEFINITION 2.4. Let $\mathcal{L}_A \to \mathcal{O}(\mathbf{A}^\bullet), \mathcal{L}_B \to \mathcal{O}(\mathbf{B}^\bullet)$ be Lagrangian structures. A *morphism of Lagrangian structures* is a commutative diagram in $D^b(U_{k+1})$

$$
\begin{array}{ccc}
\mathcal{L}_A & \longrightarrow & \mathcal{O}(\mathbf{A}^\bullet) \\
\downarrow & & \downarrow \mathcal{O}(f) \\
\mathcal{L}_B & \longrightarrow & \mathcal{O}(\mathbf{B}^\bullet)
\end{array}
$$

for some $f : \mathbf{A}^\bullet \to \mathbf{B}^\bullet$.

REMARK 2.3. Given a diagram

$$
\begin{array}{ccc}
\mathcal{L}_A & \longrightarrow & \mathcal{O}(\mathbf{A}^\bullet) \\
 & & \downarrow \mathcal{O}(f) \\
\mathcal{L}_B & \longrightarrow & \mathcal{O}(\mathbf{B}^\bullet)
\end{array}
$$

(the horizontal arrows being Lagrangian structures), there exists at most one morphism $l : \mathcal{L}_A \to \mathcal{L}_B$ which completes the above square to a commutative diagram. This can be seen as follows: Consider the long exact sequence induced on cohomology sheaves

$$\mathbf{H}^i(\mathcal{O}(\mathbf{B}^\bullet)) \to \mathbf{H}^{i+1}(\mathcal{DL}_B[n]) \to \mathbf{H}^{i+1}(\mathcal{L}_B) \to \mathbf{H}^{i+1}(\mathcal{O}(\mathbf{B}^\bullet)).$$

By definition of a Lagrangian structure, $\mathbf{H}^{i+1}(\mathcal{L}_B) \to \mathbf{H}^{i+1}(\mathcal{O}(\mathbf{B}^\bullet))$ is injective. So the sequence becomes

$$\mathbf{H}^i(\mathcal{O}(\mathbf{B}^\bullet)) \to \mathbf{H}^{i+1}(\mathcal{DL}_B[n]) \to 0.$$

Now $\mathbf{H}^i(\mathcal{O}(\mathbf{B}^\bullet)) = 0$ for $i \neq \bar{n}(k) - n$, and $\mathcal{DL}_B[n]$ is concentrated in dimension $\bar{n}(k) - n + 1$. Since $\mathcal{L}_A$ is concentrated in dimension $\bar{n}(k) - n$, it follows by [**GM83**], prop. 1.15, that

$$\mathrm{Hom}(\mathcal{L}_A, \mathcal{DL}_B[n]) = 0.$$

This shows that in the exact sequence

$$\mathrm{Hom}(\mathcal{L}_A, \mathcal{DL}_B[n]) \to \mathrm{Hom}(\mathcal{L}_A, \mathcal{L}_B) \to \mathrm{Hom}(\mathcal{L}_A, \mathcal{O}(\mathbf{B}^\bullet))$$

the map $\mathrm{Hom}(\mathcal{L}_A, \mathcal{L}_B) \to \mathrm{Hom}(\mathcal{L}_A, \mathcal{O}(\mathbf{B}^\bullet))$ is injective. Hence, $l : \mathcal{L}_A \to \mathcal{L}_B$ is unique.
□

It follows from the fact that the lifting obstruction is a functor that the composition of morphisms of Lagrangian structures is well defined. Thus Lagrangian structures form a category

$$\mathrm{Lag}(U_{k+1} - U_k).$$

The notation $U_{k+1} - U_k$ in the argument is chosen to reflect $\mathrm{supp}(\mathcal{L}) \subset U_{k+1} - U_k$ for a Lagrangian structure $\mathcal{L} \to \mathcal{O}(\mathbf{A}^\bullet)$, $\mathbf{A}^\bullet \in SD(U_k)$.

REMARK 2.4. We have to explain the terminology "Lagrangian" structure. Given a locally constant sheaf $\mathbf{H}$ on a manifold M, together with a nonsingular pairing

$$\phi : \mathbf{H} \otimes \mathbf{H} \longrightarrow \mathbb{R}_M,$$

let us call a subsheaf $\mathbf{E} \subset \mathbf{H}$ *Lagrangian* if, for every $x \in M$, the stalk $\mathbf{E}_x$ is a maximally self-annihilating (i.e. Lagrangian) subspace of $\mathbf{H}_x$ with respect to ϕ_x.

Now let $\mathbf{A}^\bullet \in SD(U_k)$, and $i : U_k \hookrightarrow U_{k+1}$, $j : U_{k+1} - U_k \hookrightarrow U_{k+1}$ be inclusions. Put $\Sigma = U_{k+1} - U_k$ and $\mathbf{H} = \mathbf{H}^s(j^* Ri_* \mathbf{A}^\bullet)$ (with $s = \bar{n}(k) - n$). Then $\mathbf{H}$ is a local system on the open stratum Σ. The self-duality isomorphism $\delta : \mathcal{D}\mathcal{O}(\mathbf{A}^\bullet)[n+1] \xrightarrow{\cong} \mathcal{O}(\mathbf{A}^\bullet)$ (lemma 2.3) induces a nonsingular pairing

$$\delta' : \mathbf{H} \otimes \mathbf{H} \longrightarrow \mathbb{R}_\Sigma.$$

We shall show that a Lagrangian subsheaf $\mathbf{E} \subset \mathbf{H}$ gives rise to a Lagrangian structure $\mathcal{L} \to \mathcal{O}(\mathbf{A}^\bullet)$. Consider the exact sequence

$$0 \longrightarrow \mathbf{E} \xrightarrow{\alpha} \mathbf{H} \xrightarrow{\beta} \mathbf{Q} \longrightarrow 0 \tag{5}$$

where $\mathbf{Q} = \mathbf{H}/\mathbf{E}$, α is the inclusion and β the quotient map. As $\mathbf{E}$ is Lagrangian, δ' induces a nonsingular pairing $\mathbf{E} \otimes \mathbf{Q} \to \mathbb{R}_\Sigma$ such that

$$\begin{array}{ccc} \mathbf{E} & \xrightarrow{\alpha} & \mathbf{H} \\ \otimes & & \otimes \\ \mathbf{Q} & \xleftarrow{\beta} & \mathbf{H} \\ \downarrow & & \downarrow \delta' \\ \mathbb{R}_\Sigma & & \mathbb{R}_\Sigma \end{array} \tag{6}$$

commutes. Let $\mathcal{L} = \mathbf{E}[-s]$, $\mathcal{Q} = \mathbf{Q}[-s]$, regarded as objects in the derived category and concentrated in dimension s. Then α and β induce unique morphisms $\mathcal{L} \to \mathcal{O}(\mathbf{A}^\bullet)$, $\mathcal{O}(\mathbf{A}^\bullet) \to \mathcal{Q}$ respectively, which we shall again denote by α, β. Now if $\mathcal{A}$ is any local system on a manifold M, then $\mathcal{D}\mathcal{A}[-\dim M] = \mathcal{A}^* \otimes or_M$, where $\mathcal{A}^*$ is the local system with stalks $\mathcal{A}^*_x = \mathrm{Hom}(\mathcal{A}_x, \mathbb{R})$ and or_M is the orientation sheaf, see [**B**$^+$**84**, V, 7.10(4)]. In the present case it follows that $\mathcal{D}\mathcal{L}[n+1] = \mathcal{L}^*$, $\mathcal{D}\mathcal{O}(\mathbf{A}^\bullet)[n+1] = \mathcal{O}(\mathbf{A}^\bullet)^*$ and $\mathcal{D}\mathcal{Q}[n+1] = \mathcal{Q}^*$, as Σ is oriented. The pairings in

(6) induce isomorphisms $\mathcal{O}^* \xrightarrow{\cong} \mathcal{O}$, $\mathcal{Q}^* \xrightarrow{\cong} \mathcal{L}$ such that

$$\begin{array}{ccc} \mathcal{L} & \xrightarrow{\alpha} & \mathcal{O} \\ \cong\uparrow & & \uparrow\cong \\ \mathcal{Q}^* & \xrightarrow{\beta^*} & \mathcal{O}^* \end{array}$$

commutes. The short exact sequence (5) gives rise to a standard triangle in the derived category and we obtain an isomorphism of triangles

$$\begin{array}{ccccccc} \mathcal{L} & \xrightarrow{\alpha} & \mathcal{O} & \xrightarrow{\beta} & \mathcal{Q} & \xrightarrow{+1} & \\ \cong\uparrow\delta_0 & & \cong\uparrow\delta & & \cong\uparrow\mathcal{D}\delta_0[n+1] & & \\ \mathcal{D}\mathcal{Q}[n+1] & \xrightarrow{\mathcal{D}\beta[n+1]} & \mathcal{D}\mathcal{O}[n+1] & \xrightarrow{(-1)^{n+1}\mathcal{D}\alpha[n+1]} & \mathcal{D}\mathcal{L}[n+1] & \xrightarrow{+1} & \end{array}$$

Using $\mathcal{D}\delta_0[n+1]$, we finally arrive at a triangle

$$\mathcal{L} \xrightarrow{\alpha} \mathcal{O} \xrightarrow{\alpha'} \mathcal{D}\mathcal{L}[n+1] \xrightarrow{+1}$$

and a diagram

$$\begin{array}{ccccccc} \mathcal{L} & \xrightarrow{\alpha} & \mathcal{O} & \xrightarrow{\alpha'} & \mathcal{D}\mathcal{L}[n+1] & \xrightarrow{+1} & \\ \| & & \cong\uparrow\delta & & \| & & \\ \mathcal{L} & \xrightarrow{\mathcal{D}\alpha'[n+1]} & \mathcal{D}\mathcal{O}[n+1] & \xrightarrow{(-1)^{n+1}\mathcal{D}\alpha[n+1]} & \mathcal{D}\mathcal{L}[n+1] & \xrightarrow{+1} & \end{array}$$

which defines the desired Lagrangian structure. Conversely, given a Lagrangian structure it is clear that it gives rise to a Lagrangian subsheaf for the relevant local system. We prefer to use the definition of Lagrangian structures as stated, because it is better adapted to the triangulation of the derived category and to Verdier duality.

4. Extracting Lagrangian Structures from Self-Dual Sheaves

We demonstrate that every self-dual complex of sheaves in $SD(X)$ has naturally associated Lagrangian structures. More precisely, we will construct a covariant functor

$$\Lambda : SD(U_{k+1}) \longrightarrow \operatorname{Lag}(U_{k+1} - U_k).$$

Suppose we are given $\mathbf{L}^\bullet \in SD(U_{k+1})$. Put

$$\mathcal{L} = \mathbf{H}^s(\mathbf{L}^\bullet)[-s] \in D^b(U_{k+1})$$

and define a morphism $\alpha : \mathbf{L}^\bullet \to \mathcal{L}$ by

$$\alpha = \Theta^{-1}(1_{\mathbf{H}^s(\mathbf{L}^\bullet)}),$$

where $\Theta : \mathrm{Hom}_{D^b(U_{k+1})}(\mathbf{L}^\bullet, \mathcal{L}) \xrightarrow{\cong} \mathrm{Hom}(\mathbf{H}^s(\mathbf{L}^\bullet), \mathbf{H}^s(\mathbf{L}^\bullet))$. We shall write $\mathbf{L}^\bullet|$ for $\mathbf{L}^\bullet|_{U_k}$. Recall the morphisms

$$^{\bar{m}}\mathbf{L}^\bullet| \xrightarrow{a} \mathbf{L}^\bullet \xrightarrow{b} {}^{\bar{n}}\mathbf{L}^\bullet|$$

defined in section 2.

THEOREM 2.3. *There is a unique morphism $\alpha' : \mathcal{L} \to {}^{\bar{m}}\boldsymbol{L}^\bullet|[1]$ such that the following is a distinguished triangle:*

(7)

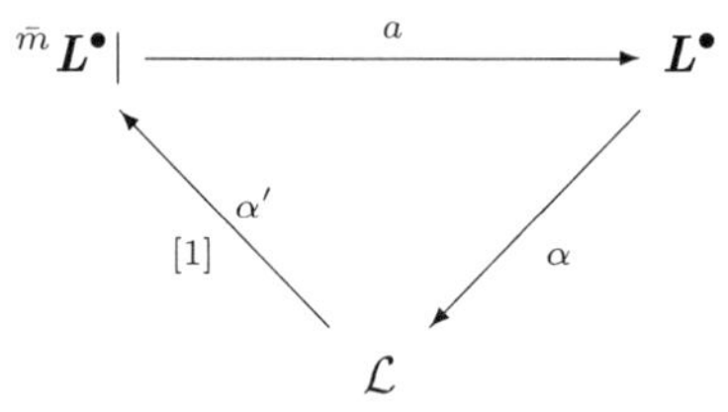

PROOF. Let

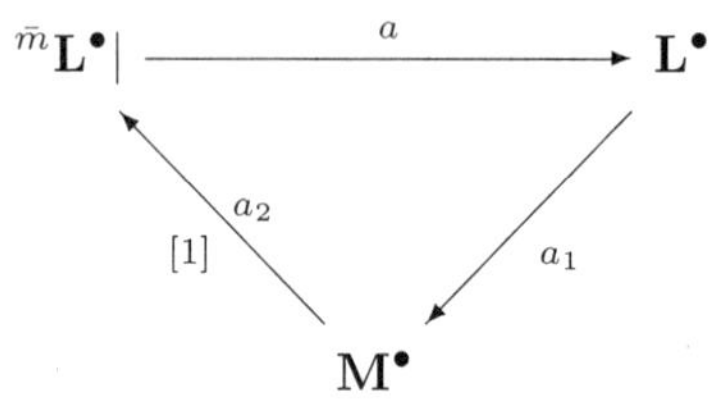

be some distinguished triangle on a. The natural map

$$\Theta : \mathrm{Hom}_{D^b(U_{k+1})}(\mathcal{L}, \mathbf{M}^\bullet) \xrightarrow{\cong} \mathrm{Hom}(\mathbf{H}^s(\mathbf{L}^\bullet), \mathbf{H}^s(\mathbf{M}^\bullet))$$

is an isomorphism because $\mathbf{H}^i(\mathbf{M}^\bullet) = 0$ for $i \neq s$. Define $\lambda : \mathcal{L} \to \mathbf{M}^\bullet$ as $\lambda = \Theta^{-1}(\mathbf{H}^s(a_1)\mathbf{H}^s(\alpha)^{-1})$. Note that λ is a quasi-isomorphism since $\mathbf{H}^i(\mathcal{L}) = \mathbf{H}^i(\mathbf{M}^\bullet) = 0$ for $i \neq s$ and $\mathbf{H}^s(\lambda)$ is an isomorphism by construction.

We obtain a commutative square

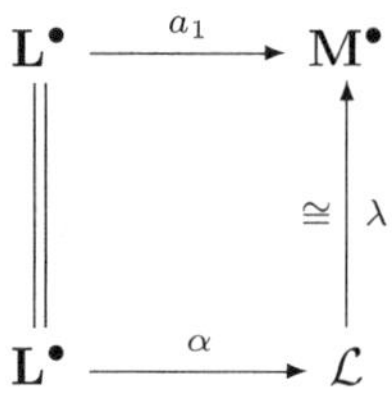

Let $\alpha' = a_2\lambda$. Then the following diagram commutes:

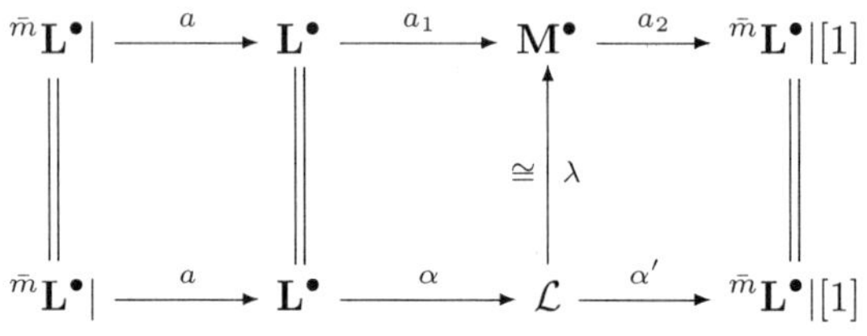

This proves that

$$ {}^{\bar{m}}\mathbf{L}^\bullet| \xrightarrow{\quad a \quad} \mathbf{L}^\bullet \xrightarrow{\quad \alpha \quad} \mathcal{L} \xrightarrow{\quad \alpha' \quad} {}^{\bar{m}}\mathbf{L}^\bullet|[1] $$

is distinguished. The uniqueness of α' follows from $\operatorname{Hom}({}^{\bar{m}}\mathbf{L}^\bullet|[1], \mathcal{L}) = 0$.

□

Next, we construct a triangle on b, essentially by dualizing the one on a given by theorem 2.3:
The dual of triangle (7) is

$$ \mathcal{D}\mathbf{L}^\bullet[n] \xrightarrow{\mathcal{D}a[n]} \mathcal{D}^{\bar{m}}\mathbf{L}^\bullet|[n] \xrightarrow{(-1)^{n+1}\mathcal{D}\alpha'[n+1]} \mathcal{D}\mathcal{L}[n+1] \xrightarrow{\mathcal{D}\alpha[n+1]} \mathcal{D}\mathbf{L}^\bullet[n+1] $$

The square

$$ \begin{array}{ccc} \mathcal{D}\mathbf{L}^\bullet[n] & \xrightarrow{\mathcal{D}a[n]} & \mathcal{D}^{\bar{m}}\mathbf{L}^\bullet|[n] \\ d \downarrow \cong & & \cong \downarrow d' \\ \mathbf{L}^\bullet & \xrightarrow{\quad b \quad} & {}^{\bar{n}}\mathbf{L}^\bullet| \end{array} $$

commutes (see the proof of theorem 2.2). Therefore, if we set

$$ \begin{aligned} \beta &= (-1)^{n+1}\mathcal{D}\alpha'[n+1] \circ d'^{-1} \\ \beta' &= d[1] \circ \mathcal{D}\alpha[n+1] \end{aligned} $$

then we get an isomorphism of triangles

$$ \begin{array}{ccccccc} \mathcal{D}\mathbf{L}^\bullet[n] & \xrightarrow{\mathcal{D}a[n]} & \mathcal{D}^{\bar{m}}\mathbf{L}^\bullet|[n] & \xrightarrow{(-1)^{n+1}\mathcal{D}\alpha'[n+1]} & \mathcal{D}\mathcal{L}[n+1] & \xrightarrow{\mathcal{D}\alpha[n+1]} & \mathcal{D}\mathbf{L}^\bullet[n+1] \\ d \downarrow \cong & & d' \downarrow \cong & & \| & & \cong \downarrow d[1] \\ \mathbf{L}^\bullet & \xrightarrow{\quad b \quad} & {}^{\bar{n}}\mathbf{L}^\bullet| & \xrightarrow{\quad \beta \quad} & \mathcal{D}\mathcal{L}[n+1] & \xrightarrow{\quad \beta' \quad} & \mathbf{L}^\bullet[1] \end{array} $$

Since ba is a factorization of the canonical morphism ${}^{\bar{m}}\mathbf{L}^\bullet| \to {}^{\bar{n}}\mathbf{L}^\bullet|$, we have a triangle

$$ (8) \qquad \begin{array}{ccc} {}^{\bar{m}}\mathbf{L}^\bullet| & \xrightarrow{\quad ba \quad} & {}^{\bar{n}}\mathbf{L}^\bullet| \\ & {}_{[1]}\nwarrow^{\omega'} \qquad \swarrow_{\omega} & \\ & \mathcal{O}(\mathbf{L}^\bullet|) & \end{array} $$

Applying the octahedral axiom to the configuration of triangles consisting of (8) and

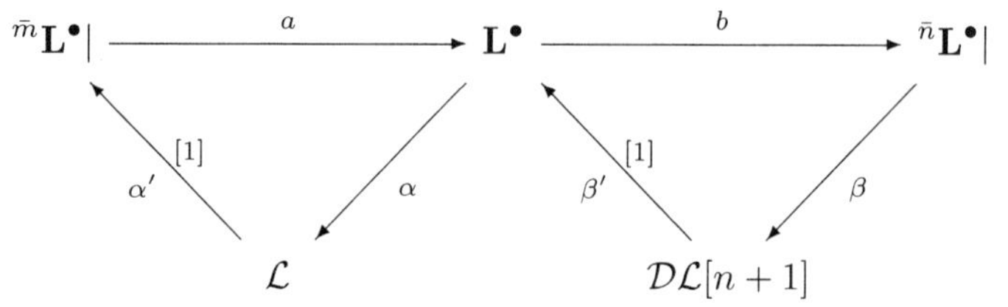

produces a distinguished triangle

(9)

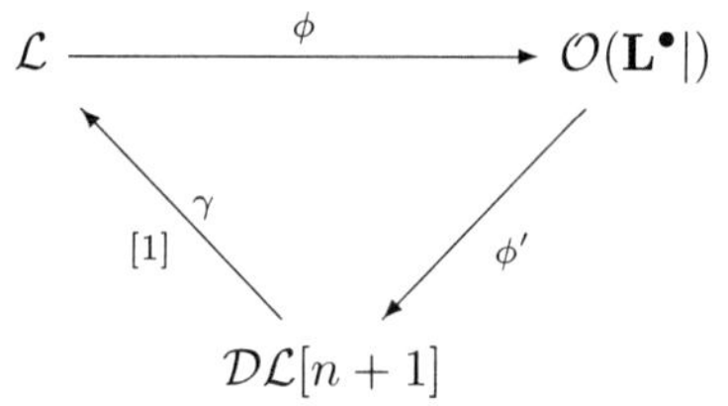

REMARK 2.5. From the commutativity relations in an octahedron, the following square commutes:

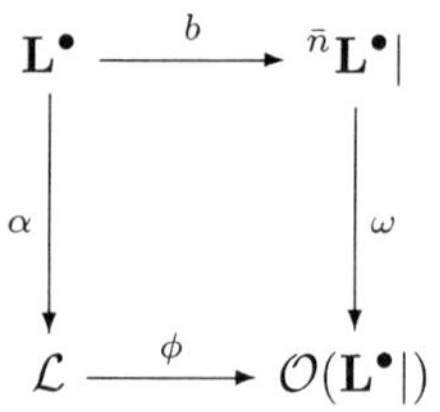

As $\mathbf{H}^s(\alpha) = 1_{\mathbf{H}^s(\mathbf{L}^\bullet)}$ and $\mathbf{H}^s(\omega) = 1_{\mathbf{H}^s(^{\bar{n}}\mathbf{L}^\bullet|)}$, we have $\mathbf{H}^s(\phi) = \mathbf{H}^s(b)$. Using the isomorphism

$$\Theta : \mathrm{Hom}_{D^b(U_{k+1})}(\mathcal{L}, \mathcal{O}(\mathbf{L}^\bullet|)) \xrightarrow{\cong} \mathrm{Hom}(\mathbf{H}^s(\mathbf{L}^\bullet), \mathbf{H}^s(^{\bar{n}}\mathbf{L}^\bullet|))$$

we identify ϕ as the map

$$\phi = \Theta^{-1}(\mathbf{H}^s(b)).$$

THEOREM 2.4. *$\mathcal{L} \xrightarrow{\phi} \mathcal{O}(\mathbf{L}^\bullet|)$ is a Lagrangian structure.*

PROOF. It has to be shown that for γ in (9)

$$\begin{aligned} &(1)\ \ \mathcal{D}\gamma[n+1] = \gamma[-1] \text{ and} \\ &(2)\ \ \delta = \delta(d|_{U_k}) \in \{d_\gamma\} \end{aligned}$$

($d : \mathcal{D}\mathbf{L}^\bullet[n] \xrightarrow{\cong} \mathbf{L}^\bullet$, $\delta : \mathcal{DO}(\mathbf{L}^\bullet|)[n+1] \xrightarrow{\cong} \mathcal{O}(\mathbf{L}^\bullet|)$.) Note that ϕ is injective on stalks since b is injective on stalks.
We verify (1): We have $\beta' = d[1] \circ \mathcal{D}\alpha[n+1]$, which implies

$$\begin{aligned} \mathcal{D}\alpha[n] &= d^{-1} \circ \beta'[-1], \\ \mathcal{D}\beta'[n+1] &= \alpha \circ d. \end{aligned}$$

From the octahedron we know $\gamma = \alpha[1] \circ \beta'$. It follows that

$$\begin{aligned} \mathcal{D}(\gamma[-1])[n] &= \mathcal{D}(\alpha \circ \beta'[-1])[n] \\ &= \mathcal{D}\beta'[n+1] \circ \mathcal{D}\alpha[n] \\ &= \alpha d \circ d^{-1}\beta'[-1] \\ &= \alpha \circ \beta'[-1] \\ &= \gamma[-1]. \end{aligned}$$

We proceed to check (2). The morphism δ is in $\{d_\gamma\}$ iff

$$\begin{array}{ccccc} \mathcal{L} & \xrightarrow{\mathcal{D}\phi'[n+1]} & \mathcal{D}\mathcal{O}(\mathbf{L}^\bullet|)[n+1] & \xrightarrow{(-1)^{n+1}\mathcal{D}\phi[n+1]} & \mathcal{D}\mathcal{L}[n+1] \\ \| & & \cong \downarrow \delta & & \| \\ \mathcal{L} & \xrightarrow{\phi} & \mathcal{O}(\mathbf{L}^\bullet|) & \xrightarrow{\phi'} & \mathcal{D}\mathcal{L}[n+1] \end{array} \tag{10}$$

commutes. We deal with the right square first. As $\mathbf{H}^i(\mathcal{L}) = \mathbf{H}^i(\mathcal{O}(\mathbf{L}^\bullet|)) = 0$ for $i \neq s$ and $\mathbf{H}^s(\phi)$ is injective, we conclude that $\mathcal{D}\mathcal{L}[n+1]$ is a complex concentrated in dimension s. Therefore

$$\mathrm{Hom}_{D^b(U_{k+1})}(\mathcal{D}\mathcal{O}(\mathbf{L}^\bullet|)[n+1], \mathcal{D}\mathcal{L}[n+1]) \xrightarrow{\cong}$$

$$\mathrm{Hom}(\mathbf{H}^s(\mathcal{D}\mathcal{O}(\mathbf{L}^\bullet|)[n+1]), \mathbf{H}^s(\mathcal{D}\mathcal{L}[n+1]))$$

is an isomorphism and it suffices to check

$$\mathbf{H}^s(\phi')\mathbf{H}^s(\delta) = \mathbf{H}^s((-1)^{n+1}\mathcal{D}\phi[n+1]).$$

Now δ satisfies $\mathbf{H}^s(\delta) = \mathbf{H}^s(\omega)\mathbf{H}^s(d')\mathbf{H}^s(\mathcal{D}\omega'[n+1])^{-1}$ (see proof of lemma 2.4), so it is enough to see

$$\mathbf{H}^s(\phi')\mathbf{H}^s(\omega)\mathbf{H}^s(d') = \mathbf{H}^s((-1)^{n+1}\mathcal{D}\phi[n+1])\mathbf{H}^s(\mathcal{D}\omega'[n+1]) \tag{11}$$

The commutativity relations in the octahedron imply

$$\begin{aligned} \beta &= \phi'\omega \\ \alpha' &= \omega'\phi. \end{aligned}$$

Hence $(-1)^{n+1}\mathcal{D}\phi[n+1] \circ \mathcal{D}\omega'[n+1] = (-1)^{n+1}\mathcal{D}\alpha'[n+1]$ and equation (11) becomes

$$\mathbf{H}^s(\beta)\mathbf{H}^s(d') = \mathbf{H}^s((-1)^{n+1}\mathcal{D}\alpha'[n+1])$$

which holds by definition of β.

We shall now verify the commutativity of the left square in the diagram (10). Again,

$$\mathrm{Hom}_{D^b(U_{k+1})}(\mathcal{L}, \mathcal{O}(\mathbf{L}^\bullet|)) \xrightarrow{\cong} \mathrm{Hom}(\mathbf{H}^s(\mathcal{L}), \mathbf{H}^s(\mathcal{O}(\mathbf{L}^\bullet|)))$$

is an isomorphism and it is enough to show

$$\mathbf{H}^s(\phi) = \mathbf{H}^s(\delta)\mathbf{H}^s(\mathcal{D}\phi'[n+1]),$$

that is

$$\mathbf{H}^s(\phi) = \mathbf{H}^s(\omega)\mathbf{H}^s(d')\mathbf{H}^s(\mathcal{D}\omega'[n+1])^{-1}\mathbf{H}^s(\mathcal{D}\phi'[n+1]) \tag{12}$$

The octahedron contains two commutative squares:

$$\begin{array}{ccc} {}^{\bar{m}}\mathbf{L}^\bullet|[1] & \xrightarrow{a[1]} & \mathbf{L}^\bullet[1] \\ \uparrow{\scriptstyle \omega'} & & \uparrow{\scriptstyle \beta'} \\ \mathcal{O}(\mathbf{L}^\bullet|) & \xrightarrow{\phi'} & \mathcal{D}\mathcal{L}[n+1] \end{array} \tag{13}$$

and

$$\begin{array}{ccc} \mathbf{L}^\bullet & \xrightarrow{\alpha} & \mathcal{L} \\ \downarrow{\scriptstyle b} & & \downarrow{\scriptstyle \phi} \\ {}^{\bar{n}}\mathbf{L}^\bullet| & \xrightarrow{\omega} & \mathcal{O}(\mathbf{L}^\bullet|) \end{array}$$

$\mathbf{H}^s(\alpha)$ is an isomorphism, so $\mathbf{H}^s(\phi) = \mathbf{H}^s(\omega)\mathbf{H}^s(b)\mathbf{H}^s(\alpha)^{-1}$ and (12) is equivalent to

$$\mathbf{H}^s(b) = \mathbf{H}^s(d')\mathbf{H}^s(\mathcal{D}\omega'[n+1])^{-1}\mathbf{H}^s(\mathcal{D}\phi'[n+1])\mathbf{H}^s(\alpha).$$

The dual of (13) is

$$\begin{array}{ccc} \mathcal{D}^{\bar{m}}\mathbf{L}^\bullet|[n] & \xleftarrow{\mathcal{D}a[n]} & \mathcal{D}\mathbf{L}^\bullet[n] \\ \downarrow{\scriptstyle \mathcal{D}\omega'[n+1]} & & \downarrow{\scriptstyle \mathcal{D}\beta'[n+1]} \\ \mathcal{D}\mathcal{O}(\mathbf{L}^\bullet|)[n+1] & \xleftarrow{\mathcal{D}\phi'[n+1]} & \mathcal{L} \end{array}$$

Since $\mathcal{D}\beta'[n+1] = \alpha d$ it follows that $\mathbf{H}^s(\mathcal{D}\beta'[n+1])$ is an isomorphism. Hence it suffices to prove

$$\mathbf{H}^s(b) = \mathbf{H}^s(d')\mathbf{H}^s(\mathcal{D}a[n])\mathbf{H}^s(\mathcal{D}\beta'[n+1])^{-1}\mathbf{H}^s(\alpha)$$

Using $\mathbf{H}^s(\mathcal{D}\beta'[n+1]) = \mathbf{H}^s(\alpha)\mathbf{H}^s(d)$, the equation reads

$$\mathbf{H}^s(b)\mathbf{H}^s(d) = \mathbf{H}^s(d')\mathbf{H}^s(\mathcal{D}a[n])$$

which follows from the commutativity of

$$\begin{array}{ccc} \mathcal{D}\mathbf{L}^\bullet[n] & \xrightarrow{\mathcal{D}a[n]} & \mathcal{D}^{\bar{m}}\mathbf{L}^\bullet|[n] \\ \downarrow{\scriptstyle d\ \cong} & & \downarrow{\scriptstyle \cong\ d'} \\ \mathbf{L}^\bullet & \xrightarrow{b} & {}^{\bar{n}}\mathbf{L}^\bullet| \end{array}$$

This places δ in $\{d_\gamma\}$ and completes the argument.

□

By the theorem we can now put

$$\boxed{\Lambda(\mathbf{L}^\bullet) := (\mathcal{L} \xrightarrow{\phi} \mathcal{O}(\mathbf{L}^\bullet|)) \in \mathrm{Lag}(U_{k+1} - U_k).}$$

Let $f : \mathbf{L}_1^\bullet \to \mathbf{L}_2^\bullet$ be a morphism, $\mathbf{L}_1^\bullet, \mathbf{L}_2^\bullet \in SD(U_{k+1})$. Then we have a diagram

$$\begin{array}{ccccc}
\mathbf{L}_1^\bullet & \xrightarrow{\cong} & \tau_{\leq \bar{n}(k)-n}\mathbf{L}_1^\bullet & \xrightarrow{\tau_{\leq}(\mathrm{adj})} & \tau_{\leq \bar{n}(k)-n} Ri_{k*}i_k^*\mathbf{L}_1^\bullet = {}^{\bar{n}}\mathbf{L}_1^\bullet| \\
\downarrow f & & \downarrow & & \downarrow {}^{\bar{n}}f| \\
\mathbf{L}_2^\bullet & \xrightarrow{\cong} & \tau_{\leq \bar{n}(k)-n}\mathbf{L}_2^\bullet & \xrightarrow{\tau_{\leq}(\mathrm{adj})} & \tau_{\leq \bar{n}(k)-n} Ri_{k*}i_k^*\mathbf{L}_2^\bullet = {}^{\bar{n}}\mathbf{L}_2^\bullet|
\end{array}$$

where the upper horizontal composition is $b_1 : \mathbf{L}_1^\bullet \to {}^{\bar{n}}\mathbf{L}_1^\bullet|$ and the lower horizontal composition is $b_2 : \mathbf{L}_2^\bullet \to {}^{\bar{n}}\mathbf{L}_2^\bullet|$. As $1 \xrightarrow{\mathrm{adj}} Ri_{k*}i_k^*$ is a natural morphism of functors, the above diagram commutes. Hence we have a commutative square on derived sheaves

$$\begin{array}{ccc}
\mathbf{H}^s(\mathbf{L}_1^\bullet) & \xrightarrow{\mathbf{H}^s(b_1)} & \mathbf{H}^s({}^{\bar{n}}\mathbf{L}_1^\bullet|) \\
\downarrow \mathbf{H}^s(f) & & \downarrow \mathbf{H}^s({}^{\bar{n}}f|) \\
\mathbf{H}^s(\mathbf{L}_2^\bullet) & \xrightarrow{\mathbf{H}^s(b_2)} & \mathbf{H}^s({}^{\bar{n}}\mathbf{L}_2^\bullet|)
\end{array} \tag{14}$$

Moreover, we have an isomorphism

$$\Theta : \mathrm{Hom}_{D^b(U_{k+1})}(\mathcal{L}_1, \mathcal{L}_2) \xrightarrow{\cong} \mathrm{Hom}(\mathbf{H}^s(\mathbf{L}_1^\bullet), \mathbf{H}^s(\mathbf{L}_2^\bullet)),$$

where $(\phi_{1,2} : \mathcal{L}_{1,2} \to \mathcal{O}(\mathbf{L}_{1,2}^\bullet|)) = \Lambda(\mathbf{L}_{1,2}^\bullet)$.
Set $l = \Theta^{-1}(\mathbf{H}^s(f))$, $l : \mathcal{L}_1 \to \mathcal{L}_2$. We need to see that

$$\begin{array}{ccc}
\mathcal{L}_1 & \xrightarrow{\phi_1} & \mathcal{O}(\mathbf{L}_1^\bullet|) \\
\downarrow l & & \downarrow \mathcal{O}(f|) \\
\mathcal{L}_2 & \xrightarrow{\phi_2} & \mathcal{O}(\mathbf{L}_2^\bullet|)
\end{array}$$

commutes. As $\mathrm{Hom}_{D^b(U_{k+1})}(\mathcal{L}_1, \mathcal{O}(\mathbf{L}_2^\bullet|)) \cong \mathrm{Hom}(\mathbf{H}^s(\mathbf{L}_1^\bullet), \mathbf{H}^s({}^{\bar{n}}\mathbf{L}_2^\bullet|))$, it suffices to show $\mathbf{H}^s(\mathcal{O}(f|)\phi_1) = \mathbf{H}^s(\phi_2 l)$. But $\mathbf{H}^s(\mathcal{O}(f|)) = \mathbf{H}^s({}^{\bar{n}}f|)$, $\mathbf{H}^s(\phi_1) = \mathbf{H}^s(b_1)$, $\mathbf{H}^s(\phi_2) = \mathbf{H}^s(b_2)$ and $\mathbf{H}^s(l) = \mathbf{H}^s(f)$. Thus by (14), we are done.

We set

$$\Lambda(f) := \begin{array}{ccc}
\mathcal{L}_1 & \xrightarrow{\phi_1} & \mathcal{O}(\mathbf{L}_1^\bullet|) \\
{}_l\downarrow & & \downarrow\mathcal{O}(f|) \\
\mathcal{L}_2 & \xrightarrow{\phi_2} & \mathcal{O}(\mathbf{L}_2^\bullet|)
\end{array}$$

so that $\Lambda(f) \in \mathrm{Hom}_{\mathrm{Lag}(U_{k+1}-U_k)}(\Lambda(\mathbf{L}_1^\bullet), \Lambda(\mathbf{L}_2^\bullet))$.

Finally, given $f : \mathbf{L}_1^\bullet \to \mathbf{L}_2^\bullet$, $g : \mathbf{L}_2^\bullet \to \mathbf{L}_3^\bullet$, we show $\Lambda(gf) = \Lambda(g)\Lambda(f)$: We have

$$\Lambda(f) = \begin{array}{ccc} \mathcal{L}_1 & \xrightarrow{\phi_1} & \mathcal{O}(\mathbf{L}_1^\bullet|) \\ {}_{l_f}\downarrow & & \downarrow\mathcal{O}(f|) \\ \mathcal{L}_2 & \xrightarrow{\phi_2} & \mathcal{O}(\mathbf{L}_2^\bullet|) \end{array}, \quad \Lambda(g) = \begin{array}{ccc} \mathcal{L}_2 & \xrightarrow{\phi_2} & \mathcal{O}(\mathbf{L}_2^\bullet|) \\ {}_{l_g}\downarrow & & \downarrow\mathcal{O}(g|) \\ \mathcal{L}_3 & \xrightarrow{\phi_3} & \mathcal{O}(\mathbf{L}_3^\bullet|) \end{array}$$

and

$$\Lambda(gf) := \begin{array}{ccc} \mathcal{L}_1 & \xrightarrow{\phi_1} & \mathcal{O}(\mathbf{L}_1^\bullet|) \\ {}_{l_{gf}}\downarrow & & \downarrow\mathcal{O}(gf|) \\ \mathcal{L}_3 & \xrightarrow{\phi_3} & \mathcal{O}(\mathbf{L}_3^\bullet|) \end{array}$$

Now $\mathbf{H}^s(l_{gf}) = \mathbf{H}^s(gf) = \mathbf{H}^s(g)\mathbf{H}^s(f) = \mathbf{H}^s(l_g)\mathbf{H}^s(l_f)$, whence $l_{gf} = l_g l_f$. Also, $\mathcal{O}(gf|) = \mathcal{O}(g|)\mathcal{O}(f|)$. This completes the construction of the functor Λ.

5. Lagrangian Structures as Building Blocks for Self-Dual Sheaves

It is the goal of this section to demonstrate that given a self-dual sheaf on U_k, a Lagrangian structure along the adjacent lower dimensional pure stratum $U_{k+1} - U_k$ naturally gives rise to a self-dual sheaf on the union U_{k+1}. The outline is as follows: We first develop a modified truncation functor $\tau_{\leq p}(\cdot,\cdot)$ which associates to a complex of sheaves $\mathbf{A}^\bullet$ and an injective sheaf map $\phi : \mathbf{E} \to \mathbf{H}^p(\mathbf{A}^\bullet)$ a complex $\tau_{\leq p}(\mathbf{A}^\bullet, \mathbf{E})$ with $\mathbf{H}^p(\tau_{\leq p}(\mathbf{A}^\bullet, \mathbf{E})) \cong \mathbf{E}$. There are canonical maps that relate $\tau_{\leq p}(\mathbf{A}^\bullet, \mathbf{E})$ to the standard truncations $\tau_{\leq p-1}\mathbf{A}^\bullet$ and $\tau_{\leq p}\mathbf{A}^\bullet$. Thus, $\tau_{\leq p}(\cdot,\cdot)$ can be viewed as an "interpolation" between $\tau_{\leq p-1}$ and $\tau_{\leq p}$, taking a specified subsheaf of the cohomology sheaf at the cut-off dimension into account. We apply this notion of truncation in constructing a functor $\boxplus$ which associates to a self-dual sheaf $\mathbf{A}^\bullet \in SD(U_k)$ and a Lagrangian structure $\phi : \mathcal{A} \to \mathcal{O}(\mathbf{A}^\bullet)$ a self-dual sheaf $\mathbf{A}^\bullet \boxplus \mathcal{A} \in SD(U_{k+1})$.

5.1. A truncation functor.

5.1. A truncation functor. Let $C(X)$ denote the abelian category of complexes of sheaves. We shall write $MS(X)$ for the morphism category associated with the category $Sh(X)$ of sheaves on X, that is, objects of $MS(X)$ are sheaf maps

$$\mathbf{E} \xrightarrow{\phi} \mathbf{F},$$

$\mathbf{E}, \mathbf{F} \in Sh(X)$, and morphisms in $MS(X)$ are commutative squares

$$\begin{array}{ccc} \mathbf{E}_1 & \xrightarrow{\phi_1} & \mathbf{F}_1 \\ {\scriptstyle g_1}\downarrow & & \downarrow{\scriptstyle g_2} \\ \mathbf{E}_2 & \xrightarrow{\phi_2} & \mathbf{F}_2 \end{array}$$

Fix an integer p. Let $C(X) \rtimes MS(X)$ denote the twisted product category whose objects are pairs

$$(\mathbf{A}^\bullet, \mathbf{E} \xrightarrow{\phi} \mathbf{F})$$

with ϕ injective and $\mathbf{F} = \mathbf{H}^p(\mathbf{A}^\bullet)$, and whose morphisms are pairs

$$(f, (g_1, g_2))$$

with $f : \mathbf{A}^\bullet \to \mathbf{B}^\bullet$ a morphism of complexes and $g_2 = \mathbf{H}^p(f)$:

$$\begin{array}{ccc} \mathbf{E}_1 & \xrightarrow{\phi_1} & \mathbf{H}^p(\mathbf{A}^\bullet) \\ \downarrow{\scriptstyle g_1} & & \downarrow{\scriptstyle g_2 = \mathbf{H}^p(f)} \\ \mathbf{E}_2 & \xrightarrow{\phi_2} & \mathbf{H}^p(\mathbf{B}^\bullet) \end{array}$$

Now consider the following construction: Let $\mathbf{A}^\bullet \in C(X)$, $\mathbf{E} \in Sh(X)$ and ϕ be an injective map $\mathbf{E} \hookrightarrow \mathbf{H}^p(\mathbf{A}^\bullet)$. If $\mathbf{A}^\bullet$ is

$$\ldots \to \mathbf{A}^{p-2} \xrightarrow{d^{p-2}} \mathbf{A}^{p-1} \xrightarrow{d^{p-1}} \mathbf{A}^p \xrightarrow{d^p} \mathbf{A}^{p+1} \to \ldots$$

then

$$\tau_{\leq p}\mathbf{A}^\bullet = \ldots \to \mathbf{A}^{p-2} \xrightarrow{d^{p-2}} \mathbf{A}^{p-1} \xrightarrow{d^{p-1}} \mathbf{ker}d^p \longrightarrow 0 \to 0 \to \ldots$$

Using the quotient map

$$\pi : \mathbf{ker}d^p \longrightarrow \frac{\mathbf{ker}d^p}{\mathbf{Im}d^{p-1}} = \mathbf{H}^p(\mathbf{A}^\bullet)$$

we define

$$\hat{\mathbf{E}} = \pi^{-1}(\phi(\mathbf{E})) \subset \mathbf{ker}d^p.$$

Let

$$\tau_{\leq p}(\mathbf{A}^\bullet, \mathbf{E}) := \ldots \to \mathbf{A}^{p-2} \xrightarrow{d^{p-2}} \mathbf{A}^{p-1} \xrightarrow{d^{p-1}} \hat{\mathbf{E}} \to 0 \to \ldots$$

(in the notation $\tau_{\leq p}(\mathbf{A}^\bullet, \mathbf{E})$, we suppress the map ϕ). This yields an assignment

$$(\mathbf{A}^\bullet, \mathbf{E} \xrightarrow{\phi} \mathbf{H}^p(\mathbf{A}^\bullet)) \mapsto \tau_{\leq p}(\mathbf{A}^\bullet, \mathbf{E}).$$

THEOREM 2.5. *$\tau_{\leq p}(\cdot,\cdot)$ is a covariant functor $C(X) \rtimes MS(X) \longrightarrow C(X)$.*

PROOF. Let

$$(f, (g, \mathbf{H}^p(f))) \in \mathrm{Hom}_{C(X)\rtimes MS(X)}((\mathbf{A}^\bullet, \mathbf{E}_1 \xrightarrow{\phi_1} \mathbf{H}^p(\mathbf{A}^\bullet)), (\mathbf{B}^\bullet, \mathbf{E}_2 \xrightarrow{\phi_2} \mathbf{H}^p(\mathbf{B}^\bullet))).$$

We shall define $\tau_{\leq p}(f,g) : \tau_{\leq p}(\mathbf{A}^\bullet, \mathbf{E}_1) \to \tau_{\leq p}(\mathbf{B}^\bullet, \mathbf{E}_2)$. Let $\hat{\mathbf{E}_1} = \pi_A^{-1}\phi_1(\mathbf{E}_1)$, $\hat{\mathbf{E}_2} = \pi_B^{-1}\phi_2(\mathbf{E}_2)$, where $\pi_A : \mathbf{ker}d_A^p \to \mathbf{H}^p(\mathbf{A}^\bullet)$, $\pi_B : \mathbf{ker}d_B^p \to \mathbf{H}^p(\mathbf{B}^\bullet)$ are the projections. In dimension p, we have $f^p : \mathbf{A}^p \to \mathbf{B}^p$. We claim that f^p restricts to $f^p| : \hat{\mathbf{E}_1} \to \hat{\mathbf{E}_2}$. Firstly, f^p restricts to a map $f^p| : \mathbf{ker}d_A^p \to \mathbf{ker}d_B^p$ as f is a morphism of complexes. Now let $x \in \hat{\mathbf{E}_1} \subset \mathbf{ker}d_A^p$. The commutativity of

$$\begin{array}{ccc} \mathbf{ker}d_A^p & \xrightarrow{f^p|} & \mathbf{ker}d_B^p \\ \downarrow{\scriptstyle \pi_A} & & \downarrow{\scriptstyle \pi_B} \\ \mathbf{H}^p(\mathbf{A}^\bullet) & \xrightarrow{\mathbf{H}^p(f)} & \mathbf{H}^p(\mathbf{B}^\bullet) \end{array}$$

means we have the identity $\pi_B(f^p(x)) = \mathbf{H}^p(f)(\pi_A(x))$ and $x \in \hat{\mathbf{E}}_1 = \pi_A^{-1}\phi_1(\mathbf{E}_1)$ implies $\pi_A(x) = \phi_1(\xi)$ for some $\xi \in \mathbf{E}_1$. Then using the commutative diagram

$$\begin{array}{ccc} \mathbf{E}_1 & \xrightarrow{\phi_1} & \mathbf{H}^p(\mathbf{A}^\bullet) \\ \Big\downarrow{g} & & \Big\downarrow{\mathbf{H}^p(f)} \\ \mathbf{E}_2 & \xrightarrow{\phi_2} & \mathbf{H}^p(\mathbf{B}^\bullet) \end{array}$$

we calculate $\mathbf{H}^p(f)(\pi_A(x)) = \mathbf{H}^p(f)(\phi_1(\xi)) = \phi_2(g(\xi)),\ g(\xi) \in \mathbf{E}_2$. Hence $\pi_B(f^p(x)) \in \phi_2\mathbf{E}_2$ and $f^p(x) \in \hat{\mathbf{E}}_2$ which proves the claim. We set $\tau_{\leq p}(f, g)$ to be

$$\begin{array}{ccccccccc} \tau_{\leq p}(\mathbf{A}^\bullet, \mathbf{E}_1) = \ldots & \longrightarrow & \mathbf{A}^{p-1} & \xrightarrow{d_A^{p-1}} & \hat{\mathbf{E}}_1 & \longrightarrow & 0 & \longrightarrow & \ldots \\ \Big\downarrow{\tau_{\leq p}(f,g)} & & \Big\downarrow{f^{p-1}} & & \Big\downarrow{f^p|} & & \Big\downarrow{0} & & \\ \tau_{\leq p}(\mathbf{B}^\bullet, \mathbf{E}_2) = \ldots & \longrightarrow & \mathbf{B}^{p-1} & \xrightarrow{d_B^{p-1}} & \hat{\mathbf{E}}_2 & \longrightarrow & 0 & \longrightarrow & \ldots \end{array}$$

Then $\tau_{\leq p}(f, g)$ is a morphism of complexes. The identities

$$\tau_{\leq p}(1_{\mathbf{A}^\bullet}, 1_{\mathbf{E}}) = 1_{\tau_{\leq p}(\mathbf{A}^\bullet, \mathbf{E})},$$
$$\tau_{\leq p}((f', g') \circ (f, g)) = \tau_{\leq p}(f', g') \circ \tau_{\leq p}(f, g)$$

clearly hold.

□

We list and prove several properties of $\tau_{\leq p}(\cdot, \cdot)$:

PROPOSITION 2.2. *There are canonical morphisms*

$$\tau_{\leq p-1} \xrightarrow{\tau^-} \tau_{\leq p}(\cdot, \cdot) \xrightarrow{\tau^+} \tau_{\leq p}$$

which

1. *factor the canonical* $\tau_{\leq p-1} \to \tau_{\leq p}$ *and*
2. *are quasi-isomorphisms in dimensions* $< p$.

(Here $\tau_{\leq p-1}$ *and* $\tau_{\leq p}$ *denote the standard truncations.)*

PROOF. We define τ^-, τ^+ to be

$$\begin{array}{llccccccc} \tau_{\leq p-1}\mathbf{A}^\bullet & = \ldots \to & \mathbf{A}^{p-2} & \to & \mathbf{ker} d^{p-1} & \to & 0 & \to 0 \to \\ \downarrow_{\tau^-} & & \| & & \cap & & \cap & \\ \tau_{\leq p}(\mathbf{A}^\bullet, \mathbf{E}) & = \ldots \to & \mathbf{A}^{p-2} & \to & \mathbf{A}^{p-1} & \to & \hat{\mathbf{E}} & \to 0 \to \\ \downarrow_{\tau^+} & & \| & & \| & & \cap & \\ \tau_{\leq p}\mathbf{A}^\bullet & = \ldots \to & \mathbf{A}^{p-2} & \to & \mathbf{A}^{p-1} & \to & \mathbf{ker} d^p & \to 0 \to \end{array}$$

This immediately implies 1., as well as 2.
Moreover, as $\tau_{\leq p}(f,g)$ is obtained by restricting f, it follows that the diagram

$$\begin{array}{ccccc}
\tau_{\leq p-1}\mathbf{A}^\bullet & \xrightarrow{\tau^-} & \tau_{\leq p}(\mathbf{A}^\bullet, \mathbf{E}_1) & \xrightarrow{\tau^+} & \tau_{\leq p}\mathbf{A}^\bullet \\
\downarrow{\scriptstyle \tau_{\leq p-1}(f)} & & \downarrow{\scriptstyle \tau_{\leq p}(f,g)} & & \downarrow{\scriptstyle \tau_{\leq p}(f)} \\
\tau_{\leq p-1}\mathbf{B}^\bullet & \xrightarrow{\tau^-} & \tau_{\leq p}(\mathbf{B}^\bullet, \mathbf{E}_2) & \xrightarrow{\tau^+} & \tau_{\leq p}\mathbf{B}^\bullet
\end{array}$$

commutes.

□

Proposition 2.3. *There exists a canonical isomorphism*

$$z : \boldsymbol{H}^p(\tau_{\leq p}(\boldsymbol{A}^\bullet, \boldsymbol{E})) \xrightarrow{\cong} \boldsymbol{E}$$

such that

$$\begin{array}{ccc}
\boldsymbol{H}^p(\tau_{\leq p}(\boldsymbol{A}^\bullet, \boldsymbol{E})) & \xrightarrow{\boldsymbol{H}^p(\tau^+)} & \boldsymbol{H}^p(\tau_{\leq p}\boldsymbol{A}^\bullet) \\
{\scriptstyle z}\downarrow{\scriptstyle \cong} & & \| \\
\boldsymbol{E} & \xrightarrow{\phi} & \boldsymbol{H}^p(\boldsymbol{A}^\bullet)
\end{array}$$

commutes.

Proof. Let z be the composition

$$z : \mathbf{H}^p(\tau_{\leq p}(\mathbf{A}^\bullet, \mathbf{E})) = \frac{\pi^{-1}\phi(\mathbf{E})}{\mathbf{Im} d^{p-1}} = \pi(\pi^{-1}\phi(\mathbf{E})) = \phi(\mathbf{E}) \xrightarrow{\phi|^{-1}} \mathbf{E}.$$

Then the diagram mentioned in the proposition is

$$\begin{array}{ccc}
\mathbf{H}^p(\tau_{\leq p}(\mathbf{A}^\bullet, \mathbf{E})) & \xrightarrow{\mathbf{H}^p(\tau^+)} & \mathbf{H}^p(\tau_{\leq p}\mathbf{A}^\bullet) \\
\| & & \| \\
\dfrac{\pi^{-1}\phi(\mathbf{E})}{\mathbf{Im} d^{p-1}} & \xrightarrow{\text{incl}} & \dfrac{\mathbf{ker} d^p}{\mathbf{Im} d^{p-1}} \\
\| & & \| \\
\phi(\mathbf{E}) & \xrightarrow{\text{incl}} & \mathbf{H}^p(\mathbf{A}^\bullet) \\
{\scriptstyle \cong}\downarrow{\scriptstyle \phi|^{-1}} & & \| \\
\mathbf{E} & \xrightarrow{\phi} & \mathbf{H}^p(\mathbf{A}^\bullet)
\end{array}$$

which commutes. Given a morphism (f, g) with

$$\begin{array}{ccc} \mathbf{E}_1 & \xrightarrow{\phi_1} & \mathbf{H}^p(\mathbf{A}^\bullet) \\ \Big\downarrow g & & \Big\downarrow \mathbf{H}^p(f) \\ \mathbf{E}_2 & \xrightarrow{\phi_2} & \mathbf{H}^p(\mathbf{B}^\bullet) \end{array}$$

commutative, the map $\mathbf{H}^p(f)$ restricts so that we have a commutative square

$$\begin{array}{ccc} \phi_1(\mathbf{E}_1) & \xrightarrow{\mathbf{H}^p(f)|} & \phi_2(\mathbf{E}_2) \\ \Big\downarrow \phi_1|^{-1} & & \Big\downarrow \phi_2|^{-1} \\ \mathbf{E}_1 & \xrightarrow{g} & \mathbf{E}_2 \end{array}$$

Therefore,

$$\begin{array}{ccc} \mathbf{H}^p(\tau_{\leq p}(\mathbf{A}^\bullet, \mathbf{E}_1)) & \xrightarrow{\mathbf{H}^p(\tau_{\leq p}(f,g))} & \mathbf{H}^p(\tau_{\leq p}(\mathbf{B}^\bullet, \mathbf{E}_2)) \\ z \Big\downarrow \cong & & \cong \Big\downarrow z \\ \mathbf{E}_1 & \xrightarrow{g} & \mathbf{E}_2 \end{array}$$

commutes and z is canonical.

□

PROPOSITION 2.4. *Let* $(f, (g, \boldsymbol{H}^p(f))) \in \mathrm{Hom}_{C(X) \rtimes MS(X)}((\boldsymbol{A}^\bullet, \boldsymbol{E}_1 \xrightarrow{\phi_1} \boldsymbol{H}^p(\boldsymbol{A}^\bullet)),$ $(\boldsymbol{B}^\bullet, \boldsymbol{E}_2 \xrightarrow{\phi_2} \boldsymbol{H}^p(\boldsymbol{B}^\bullet)))$ *be a morphism such that* $f : \boldsymbol{A}^\bullet \to \boldsymbol{B}^\bullet$ *is homotopic to zero. Then* $\tau_{\leq p}(f, g)$ *is homotopic to zero.*

PROOF. Let $\{s^i\}$, $s^i : \mathbf{A}^i \to \mathbf{B}^{i-1}$, be a nullhomotopy:

$$f^i = s^{i+1} d_A^i + d_B^{i-1} s^i.$$

Define $\{t^i\}$, $t^i : \tau_{\leq p}(\mathbf{A}^\bullet, \mathbf{E}_1)^i \to \tau_{\leq p}(\mathbf{B}^\bullet, \mathbf{E}_2)^{i-1}$ by

$$t^i = \begin{cases} s^i, & \text{for } i \leq p-1 \\ s^p|_{\hat{\mathbf{E}}_1}, & \text{for } i = p \\ 0, & \text{for } i \geq p+1 \end{cases}$$

We check that $\{t^i\}$ is a nullhomotopy for $\tau_{\leq p}(f, g)$:

$$\begin{aligned} \tau_{\leq p}(f,g)^p = f^p|_{\hat{\mathbf{E}}_1} &= f^p|_{\hat{\mathbf{E}}_1} - s^{p+1} d_A^p|_{\hat{\mathbf{E}}_1} = d_B^{p-1} s^p|_{\hat{\mathbf{E}}_1} \\ &= t^{p+1} 0 + d_B^{p-1} t^p \end{aligned}$$

and the equations in all other dimensions are clear.

□

Let $K(X)$ be the homotopy category of complexes of sheaves on X. The previous proposition shows that $\tau_{\leq p}(\cdot,\cdot)$ is a well defined functor

$$\tau_{\leq p}(\cdot,\cdot) : K(X) \rtimes MS(X) \longrightarrow K(X)$$

PROPOSITION 2.5. *Let* $(f,g) \in \mathrm{Hom}_{C(X)\rtimes MS(X)}((\boldsymbol{A}^\bullet, \boldsymbol{E}_1 \overset{\phi_1}{\to} \boldsymbol{H}^p(\boldsymbol{A}^\bullet)), (\boldsymbol{B}^\bullet, \boldsymbol{E}_2 \overset{\phi_2}{\to} \boldsymbol{H}^p(\boldsymbol{B}^\bullet)))$ *be a morphism such that* $f : \boldsymbol{A}^\bullet \to \boldsymbol{B}^\bullet$ *is a quasi-isomorphism and* g *is an isomorphism. Then* $\tau_{\leq p}(f,g)$ *is a quasi-isomorphism.*

PROOF. For $i < p$, $\mathbf{H}^i(\tau_{\leq p}(f,g)) = \mathbf{H}^i(f)$ which is an isomorphism. For $i > p$, $\mathbf{H}^i(\tau_{\leq p}(\mathbf{A}^\bullet, \mathbf{E}_1)) = \mathbf{H}^i(\tau_{\leq p}(\mathbf{B}^\bullet, \mathbf{E}_2)) = 0$. It remains to see that $\mathbf{H}^p(\tau_{\leq p}(f,g))$ is an isomorphism. Recall the canonical map z given by proposition 2.3. By naturality, we have the commutative diagram

$$\begin{array}{ccc}
\mathbf{H}^p(\tau_{\leq p}(\mathbf{A}^\bullet, \mathbf{E}_1)) & \xrightarrow{\mathbf{H}^p(\tau_{\leq p}(f,g))} & \mathbf{H}^p(\tau_{\leq p}(\mathbf{B}^\bullet, \mathbf{E}_2)) \\
{\scriptstyle z}\downarrow{\scriptstyle \cong} & & {\scriptstyle \cong}\downarrow{\scriptstyle z} \\
\mathbf{E}_1 & \xrightarrow[\cong]{g} & \mathbf{E}_2
\end{array}$$

Hence $\mathbf{H}^p(\tau_{\leq p}(f,g))$ is an isomorphism.

□

Proposition 2.5 implies that $\tau_{\leq p}(\cdot,\cdot)$ is in fact a well defined functor on the derived category:

$$\tau_{\leq p}(\cdot,\cdot) : D(X) \rtimes MS(X) \longrightarrow D(X)$$

REMARK 2.6. If $Q : C(X) \to D(X)$ is the canonical quotient functor, we shall still write τ^-, τ^+ for the morphisms $Q(\tau^-), Q(\tau^+)$ in the derived category (for τ^-, τ^+ see proposition 2.2).

5.2. The construction of ⊞. Let $SD(U_k) \rtimes \mathrm{Lag}(U_{k+1} - U_k)$ denote the twisted product of categories whose objects are pairs

$$(\mathbf{A}^\bullet, \mathcal{A} \xrightarrow{\phi} \mathcal{O}(\mathbf{A}^\bullet)),$$

$\mathbf{A}^\bullet \in SD(U_k)$, $\phi \in \mathrm{Lag}(U_{k+1} - U_k)$, and whose morphisms are pairs with first component a morphism $f \in \mathrm{Hom}_{D^b(U_k)}(\mathbf{A}^\bullet, \mathbf{B}^\bullet)$ and second component a commutative square

$$\begin{array}{ccc}
\mathcal{A} & \xrightarrow{\phi_A} & \mathcal{O}(\mathbf{A}^\bullet) \\
{\scriptstyle l}\downarrow & & \downarrow{\scriptstyle \mathcal{O}(f)} \\
\mathcal{B} & \xrightarrow{\phi_B} & \mathcal{O}(\mathbf{B}^\bullet)
\end{array}$$

(We will usually denote such a pair simply by (f,l).)
The objective of the present section is to construct a functor

$$\boxplus : SD(U_k) \rtimes \mathrm{Lag}(U_{k+1} - U_k) \longrightarrow SD(U_{k+1})$$

and to develop its properties. This will explain how one can successively use Lagrangian structures to construct objects in $SD(X)$.

Suppose we are given $\mathbf{A}^\bullet \in SD(U_k)$ and $(\mathcal{A} \xrightarrow{\phi} \mathcal{O}(\mathbf{A}^\bullet)) \in \mathrm{Lag}(U_{k+1} - U_k)$. Define

$$\boxed{\mathbf{A}^\bullet \boxplus \mathcal{A} := \tau_{\leq s}(Ri_{k*}\mathbf{A}^\bullet, \mathbf{H}^s(\mathcal{A}) \xrightarrow{\mathbf{H}^s(\phi)} \mathbf{H}^s(Ri_{k*}\mathbf{A}^\bullet))}$$

(In the notation $\mathbf{A}^\bullet \boxplus \mathcal{A}$, we suppress the map ϕ.)

Also, given $f : \mathbf{A}^\bullet \to \mathbf{B}^\bullet$ and

$$\begin{array}{ccc} \mathcal{A} & \xrightarrow{\phi_A} & \mathcal{O}(\mathbf{A}^\bullet) \\ {\scriptstyle l}\downarrow & & \downarrow{\scriptstyle \mathcal{O}(f)} \\ \mathcal{B} & \xrightarrow{\phi_B} & \mathcal{O}(\mathbf{B}^\bullet) \end{array}$$

let

$$\boxed{f \boxplus l := \tau_{\leq s}(Ri_{k*}f, \mathbf{H}^s(l))}$$

where we wrote $\mathbf{H}^s(l)$ to abbreviate the commutative square

$$\begin{array}{ccc} \mathbf{H}^s(\mathcal{A}) & \longrightarrow & \mathbf{H}^s(\mathcal{O}(\mathbf{A}^\bullet)) \\ {\scriptstyle \mathbf{H}^s(l)}\downarrow & & \downarrow{\scriptstyle \mathbf{H}^s(Ri_{k*}f)} \\ \mathbf{H}^s(\mathcal{B}) & \longrightarrow & \mathbf{H}^s(\mathcal{O}(\mathbf{B}^\bullet)) \end{array}$$

We proceed to discuss several of the properties of $\boxplus$. By proposition 2.2, we have canonical morphisms

$$\tau_{\leq s-1}Ri_{k*}\mathbf{A}^\bullet \xrightarrow{\tau^-} \tau_{\leq s}(Ri_{k*}\mathbf{A}^\bullet, \mathbf{H}^s(\phi)) \xrightarrow{\tau^+} \tau_{\leq s}Ri_{k*}\mathbf{A}^\bullet$$

Therefore, associated to $\boxplus$ we have canonical morphisms

$${}^{\bar{m}}\mathbf{A}^\bullet \xrightarrow{\tau^-} \mathbf{A}^\bullet \boxplus \mathcal{A} \xrightarrow{\tau^+} {}^{\bar{n}}\mathbf{A}^\bullet$$

which factor the canonical ${}^{\bar{m}}\mathbf{A}^\bullet \to {}^{\bar{n}}\mathbf{A}^\bullet$. By proposition 2.3, there is a canonical isomorphism

$$z : \mathbf{H}^s(\tau_{\leq s}(Ri_{k*}\mathbf{A}^\bullet, \mathbf{H}^s(\phi))) \xrightarrow{\cong} \mathbf{H}^s(\mathcal{A}).$$

Using the isomorphism

$$\Theta : \mathrm{Hom}_{D^b(U_{k+1})}(\tau_{\leq s}(Ri_{k*}\mathbf{A}^\bullet, \mathbf{H}^s(\phi)), \mathcal{A}) \xrightarrow{\cong}$$

$$\mathrm{Hom}(\mathbf{H}^s(\tau_{\leq s}(Ri_{k*}\mathbf{A}^\bullet, \mathbf{H}^s(\mathcal{A}))), \mathbf{H}^s(\mathcal{A}))$$

we obtain a morphism $\Theta^{-1}(z) : \mathbf{A}^\bullet \boxplus \mathcal{A} \to \mathcal{A}$, which we will again denote by z. By proposition 2.3, z has moreover the property that

$$\begin{array}{ccc} \mathbf{H}^s(\mathbf{A}^\bullet \boxplus \mathcal{A}) & \xrightarrow{\mathbf{H}^s(\tau^+)} & \mathbf{H}^s({}^{\bar{n}}\mathbf{A}^\bullet) \\ \mathbf{H}^s(z) \Big\downarrow \cong & & \cong \Big\downarrow \mathbf{H}^s(\omega) \\ \mathbf{H}^s(\mathcal{A}) & \xrightarrow{\mathbf{H}^s(\phi)} & \mathbf{H}^s(\mathcal{O}(\mathbf{A}^\bullet)) \end{array}$$

commutes (recall $\mathbf{H}^s(\omega) = 1_{\mathbf{H}^s(Ri_{k*}\mathbf{A}^\bullet)}$). As the natural map

$$\operatorname{Hom}_{D^b(U_{k+1})}(\mathbf{A}^\bullet \boxplus \mathcal{A}, \mathcal{O}(\mathbf{A}^\bullet)) \xrightarrow{\cong} \operatorname{Hom}(\mathbf{H}^s(\mathbf{A}^\bullet \boxplus \mathcal{A}), \mathbf{H}^s(\mathcal{O}(\mathbf{A}^\bullet)))$$

is an isomorphism, $\mathbf{H}^s(\omega\tau^+) = \mathbf{H}^s(\phi z)$ implies the commutativity of the diagram

$$\begin{array}{ccc} \mathbf{A}^\bullet \boxplus \mathcal{A} & \xrightarrow{\tau^+} & {}^{\bar{n}}\mathbf{A}^\bullet \\ z \Big\downarrow & & \Big\downarrow \omega \\ \mathcal{A} & \xrightarrow{\phi} & \mathcal{O}(\mathbf{A}^\bullet) \end{array} \tag{15}$$

It is a very attractive feature of $\mathbf{A}^\bullet \boxplus \mathcal{A}$ that one can canonically associate to it a distinguished triangle in which all morphisms are identified as already known ones. Recall our notation $\omega' : \mathcal{O}(\mathbf{A}^\bullet) \to {}^{\bar{m}}\mathbf{A}^\bullet[1]$ for the map raising degrees in the lifting obstruction triangle, theorem 2.1.

THEOREM 2.6. *Let* $\mathbf{A}^\bullet \in SD(U_k)$, $(\mathcal{A} \xrightarrow{\phi} \mathcal{O}(\mathbf{A}^\bullet)) \in \operatorname{Lag}(U_{k+1} - U_k)$. *The following is a distinguished triangle:*

$$\begin{array}{ccc} \mathcal{A} & \xrightarrow{\omega'\phi} & {}^{\bar{m}}\mathbf{A}^\bullet[1] \\ {\scriptstyle [1]}\nwarrow {\scriptstyle -z[1]} & & \swarrow {\scriptstyle -\tau^-[1]} \\ & (\mathbf{A}^\bullet \boxplus \mathcal{A})[1] & \end{array}$$

PROOF. The argument consists of two steps. In the first step we show that for some $t : \mathcal{A} \to {}^{\bar{m}}\mathbf{A}^\bullet[1]$ the triangle

$$\begin{array}{ccc} \mathcal{A} & \xrightarrow{t} & {}^{\bar{m}}\mathbf{A}^\bullet[1] \\ {\scriptstyle [1]}\nwarrow {\scriptstyle -z[1]} & & \swarrow {\scriptstyle -\tau^-[1]} \\ & (\mathbf{A}^\bullet \boxplus \mathcal{A})[1] & \end{array}$$

is distinguished. The second step identifies t as $\omega'\phi$. Let

(16)

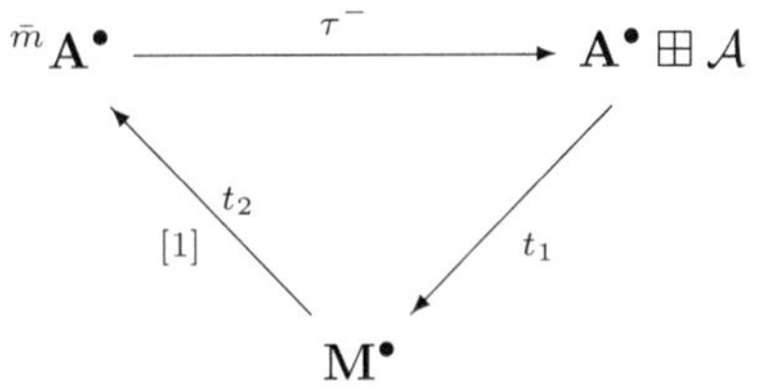

be some distinguished triangle on τ^-. We have an isomorphism

$$\Theta : \mathrm{Hom}_{D^b(U_{k+1})}(\mathbf{A}^\bullet \boxplus \mathcal{A}, \mathcal{A}) \xrightarrow{\cong} \mathrm{Hom}(\mathbf{H}^s(\mathbf{A}^\bullet \boxplus \mathcal{A}), \mathbf{H}^s(\mathcal{A})).$$

Using that $\mathbf{H}^i(\tau^-)$ is an isomorphism for $i < s$ and $\mathbf{H}^i({}^{\bar{m}}\mathbf{A}^\bullet) = 0$ for $i \geq s$, $\mathbf{H}^i(\mathbf{A}^\bullet \boxplus \mathcal{A}) = 0$ for $i > s$, the long exact sequence on derived sheaves induced by (16) shows that $\mathbf{M}^\bullet$ is a complex concentrated in dimension s. Hence the following is an isomorphism

$$\tilde{\Theta} : \mathrm{Hom}_{D^b(U_{k+1})}(\mathbf{M}^\bullet, \mathcal{A}) \xrightarrow{\cong} \mathrm{Hom}(\mathbf{H}^s(\mathbf{M}^\bullet), \mathbf{H}^s(\mathcal{A})).$$

Again looking at the long exact sequence of derived sheaves, we see that $\mathbf{H}^s(t_1)$ is an isomorphism. Set

$$\lambda = \tilde{\Theta}^{-1}(\mathbf{H}^s(z)\mathbf{H}^s(t_1)^{-1}), \lambda : \mathbf{M}^\bullet \to \mathcal{A},$$

so that $\mathbf{H}^s(\lambda t_1) = \mathbf{H}^s(z)$, and hence $\lambda t_1 = z$ by the injectivity of Θ. Note that λ is a quasi-isomorphism as $\mathbf{H}^s(\lambda)$ is a quasi-isomorphism and $\mathbf{H}^i(\mathbf{M}^\bullet) = \mathbf{H}^i(\mathcal{A}) = 0$ for $i \neq s$.
It follows that with $t = t_2\lambda^{-1}$, we have a commutative diagram

$$\begin{array}{ccccccc}
{}^{\bar{m}}\mathbf{A}^\bullet & \xrightarrow{\tau^-} & \mathbf{A}^\bullet \boxplus \mathcal{A} & \xrightarrow{t_1} & \mathbf{M}^\bullet & \xrightarrow{t_2} & {}^{\bar{m}}\mathbf{A}^\bullet[1] \\
\| & & \| & & \cong \downarrow \lambda & & \| \\
{}^{\bar{m}}\mathbf{A}^\bullet & \xrightarrow{\tau^-} & \mathbf{A}^\bullet \boxplus \mathcal{A} & \xrightarrow{z} & \mathcal{A} & \xrightarrow{t} & {}^{\bar{m}}\mathbf{A}^\bullet[1]
\end{array}$$

Since the upper row is a distinguished triangle and all vertical maps are isomorphisms, the lower row is distinguished as well. Now turning the lower triangle two times one obtains

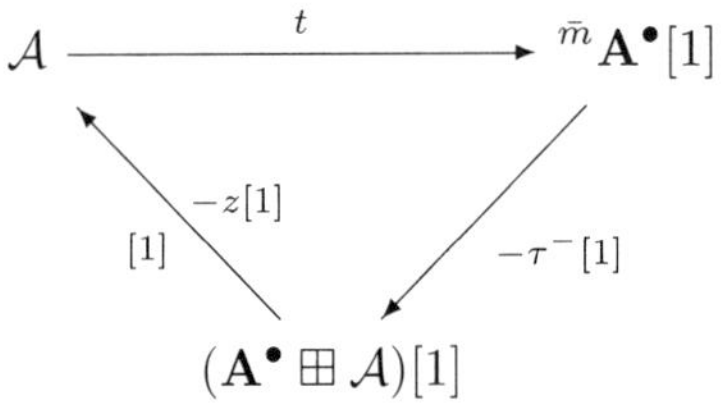

This completes step one. We shall now prove $t = \omega'\phi$. The plan is as follows: First, we show that t has a unique lift ψ to $\mathcal{O}(\mathbf{A}^\bullet)$ with respect to ω' :

$$\begin{array}{ccc} & & \mathcal{O}(\mathbf{A}^\bullet) \\ & \overset{\psi}{\nearrow} & \big\downarrow \omega' \\ \mathcal{A} & \xrightarrow{\ t\ } & {}^{\bar m}\mathbf{A}^\bullet[1] \end{array}$$

Second, we prove $\psi = \phi$.
Let c be the canonical morphism $c : {}^{\bar m}\mathbf{A}^\bullet \to {}^{\bar n}\mathbf{A}^\bullet$. The commutative square

$$\begin{array}{ccc} {}^{\bar m}\mathbf{A}^\bullet[1] & \xrightarrow{-\tau^-[1]} & (\mathbf{A}^\bullet \boxplus \mathcal{A})[1] \\ \Big\| & & \big\downarrow \tau^+[1] \\ {}^{\bar m}\mathbf{A}^\bullet[1] & \xrightarrow{-c[1]} & {}^{\bar n}\mathbf{A}^\bullet[1] \end{array}$$

can be embedded in a morphism of triangles

$$\begin{array}{ccccccc} \mathcal{A} & \xrightarrow{t} & {}^{\bar m}\mathbf{A}^\bullet[1] & \xrightarrow{-\tau^-[1]} & (\mathbf{A}^\bullet \boxplus \mathcal{A})[1] & \xrightarrow{-z[1]} & \mathcal{A}[1] \\ \big\downarrow & & \Big\| & & \big\downarrow \tau^+[1] & & \big\downarrow \\ \mathcal{O}(\mathbf{A}^\bullet) & \xrightarrow{\omega'} & {}^{\bar m}\mathbf{A}^\bullet[1] & \xrightarrow{-c[1]} & {}^{\bar n}\mathbf{A}^\bullet[1] & \xrightarrow{-\omega[1]} & \mathcal{O}(\mathbf{A}^\bullet)[1] \end{array} \tag{17}$$

Therefore, $c[1] \circ t = (\tau^+[1] \circ \tau^-[1]) \circ t = \tau^+[1] \circ 0 = 0$.
Let $x \in X_{n-l} - X_{n-l-1}$, $l \geq 2$. We compute

$$\mathbf{H}^i(j_x^! \, {}^{\bar n}\mathbf{A}^\bullet) \cong \mathbf{H}^{i+n}(\mathcal{D}(j_x^{*\bar m}\mathbf{A}^\bullet)) \cong \mathrm{Hom}(\mathbf{H}^{-i-n}({}^{\bar m}\mathbf{A}^\bullet)_x, \mathbb{R}) = 0$$

for $-i-n > \bar n(l) - l$, i.e. for $i \leq \bar n(l) - l$. Since also $\mathbf{H}^i(\mathcal{A}|_{U_{l+1}}) = 0$ for $i > \bar n(l) - n$, we may apply [**CS91**], (1.3) to conclude that restriction to U_2 is an injection

$$\mathrm{Hom}_{D^b(U_{k+1})}(\mathcal{A}, {}^{\bar n}\mathbf{A}^\bullet) \hookrightarrow \mathrm{Hom}_{D^b(U_2)}(\mathcal{A}|_{U_2}, {}^{\bar n}\mathbf{A}^\bullet|_{U_2})$$

But $\mathcal{A}|_{U_2} = 0$, so

$$\mathrm{Hom}_{D^b(U_{k+1})}(\mathcal{A}, {}^{\bar n}\mathbf{A}^\bullet) = 0.$$

Consider the long exact sequence associated to the lifting obstruction triangle

$$0 = \mathrm{Hom}(\mathcal{A}, {}^{\bar n}\mathbf{A}^\bullet) \to \mathrm{Hom}(\mathcal{A}, \mathcal{O}(\mathbf{A}^\bullet)) \overset{\omega'\circ}{\to} \mathrm{Hom}(\mathcal{A}, {}^{\bar m}\mathbf{A}^\bullet[1]) \overset{c[1]\circ}{\to} \mathrm{Hom}(\mathcal{A}, {}^{\bar n}\mathbf{A}^\bullet[1])$$

As $c[1] \circ t = 0$, there exists $\psi \in \mathrm{Hom}(\mathcal{A}, \mathcal{O}(\mathbf{A}^\bullet))$ such that $t = \omega'\psi$, and ψ is unique as composition with ω' is injective.

Finally, we prove that $\psi = \phi$. Again consider (17):

$$\begin{array}{ccccccc}
\mathcal{A} & \xrightarrow{\omega'\psi} & {}^{\bar m}\mathbf{A}^\bullet[1] & \xrightarrow{-\tau^-[1]} & (\mathbf{A}^\bullet \boxplus \mathcal{A})[1] & \xrightarrow{-z[1]} & \mathcal{A}[1] \\
\downarrow{\scriptstyle \lambda} & & \| & & \downarrow{\scriptstyle \tau^+[1]} & & \downarrow{\scriptstyle \lambda[1]} \\
\mathcal{O}(\mathbf{A}^\bullet) & \xrightarrow{\omega'} & {}^{\bar m}\mathbf{A}^\bullet[1] & \xrightarrow{-c[1]} & {}^{\bar n}\mathbf{A}^\bullet[1] & \xrightarrow{-\omega[1]} & \mathcal{O}(\mathbf{A}^\bullet)[1]
\end{array}$$

From the left square we see $\omega'\lambda = \omega'\psi = t$. Since the lift is unique, one has $\lambda = \psi$. Then from the right square we have $\lambda z = \omega\tau^+$. Using (15), we calculate on cohomology:

$$\begin{aligned}
\mathbf{H}^s(\lambda) &= \mathbf{H}^s(\omega)\mathbf{H}^s(\tau^+)\mathbf{H}^s(z)^{-1} \\
&= \mathbf{H}^s(\phi)\mathbf{H}^s(z)\mathbf{H}^s(z)^{-1} \\
&= \mathbf{H}^s(\phi)
\end{aligned}$$

The injectivity of $\mathrm{Hom}_{D^b(U_{k+1})}(\mathcal{A}, \mathcal{O}(\mathbf{A}^\bullet)) \xrightarrow{\cong} \mathrm{Hom}(\mathbf{H}^s(\mathcal{A}), \mathbf{H}^s(\mathcal{O}(\mathbf{A}^\bullet)))$ implies $\lambda = \phi$, so $\psi = \phi$.

□

Next, we show that one can canonically identify $(\mathbf{A}^\bullet \boxplus \mathcal{A})|_{U_k}$ with $\mathbf{A}^\bullet$.

PROPOSITION 2.6. *Let* $\mathbf{A}^\bullet \in SD(U_k)$, $(\mathcal{A} \xrightarrow{\phi} \mathcal{O}(\mathbf{A}^\bullet)) \in \mathrm{Lag}(U_{k+1} - U_k)$. *Then there is a canonical isomorphism*

$$\rho : \mathbf{A}^\bullet \xrightarrow{\cong} (\mathbf{A}^\bullet \boxplus \mathcal{A})|_{U_k}$$

PROOF. Firstly, as $\mathbf{A}^\bullet \cong \tau_{\leq s-1}\mathbf{A}^\bullet$ and $i_k^* Ri_{k*} \cong 1$, there is a canonical isomorphism

$$\mathbf{A}^\bullet \cong i_k^{*\bar m}\mathbf{A}^\bullet.$$

Secondly, apply i_k^* to the $\boxplus$-triangle given by theorem 2.6:

$$\begin{array}{ccc}
i_k^{*\bar m}\mathbf{A}^\bullet & \xrightarrow{i_k^*\tau^-} & i_k^*(\mathbf{A}^\bullet \boxplus \mathcal{A}) \\
& {\scriptstyle [1]}\nwarrow \quad \swarrow & \\
& i_k^*\mathcal{A} &
\end{array}$$

As $i_k^*\mathcal{A} = 0$, $i_k^*\tau^-$ is an isomorphism. We define ρ as the composition

$$\rho : \mathbf{A}^\bullet \cong i_k^{*\bar m}\mathbf{A}^\bullet \xrightarrow{i_k^*\tau^-} (\mathbf{A}^\bullet \boxplus \mathcal{A})|_{U_k}.$$

□

The following theorem completes the construction of $\boxplus$.

THEOREM 2.7. $\boxplus$ *is a covariant functor*

$$\boxplus : SD(U_k) \rtimes \mathrm{Lag}(U_{k+1} - U_k) \longrightarrow SD(U_{k+1})$$

PROOF. Given $\mathbf{A}^\bullet \in SD(U_k)$, $(\mathcal{A} \xrightarrow{\phi} \mathcal{O}(\mathbf{A}^\bullet)) \in \mathrm{Lag}(U_{k+1} - U_k)$, we have to check that $\mathbf{A}^\bullet \boxplus \mathcal{A} \in SD(U_{k+1})$. Axioms (**SD1**) — (**SD3**) being clearly satisfied, it remains to prove (**SD4**). That is, we have to construct a self-duality isomorphism

$$d : \mathcal{D}(\mathbf{A}^\bullet \boxplus \mathcal{A})[n] \xrightarrow{\cong} \mathbf{A}^\bullet \boxplus \mathcal{A}$$

such that $\mathcal{D}d[n] = d$. Put $\mathbf{L}^\bullet = \mathbf{A}^\bullet \boxplus \mathcal{A}$.
Applying the octahedral axiom to the configuration of triangles

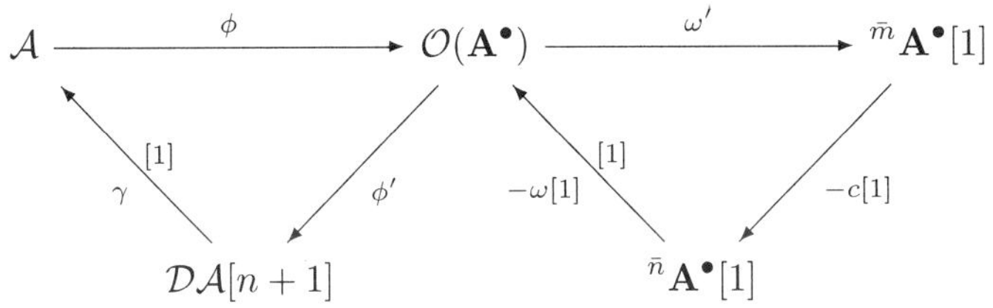

and

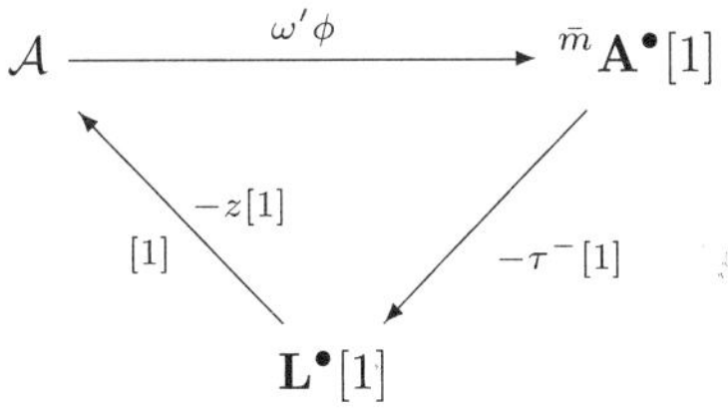

(these triangles are the triangles associated to a Lagrangian structure, see definition 2.3, the lifting obstruction triangle, see theorem 2.1, and the $\boxplus$-triangle, see theorem 2.6, respectively) produces a triangle

(18)

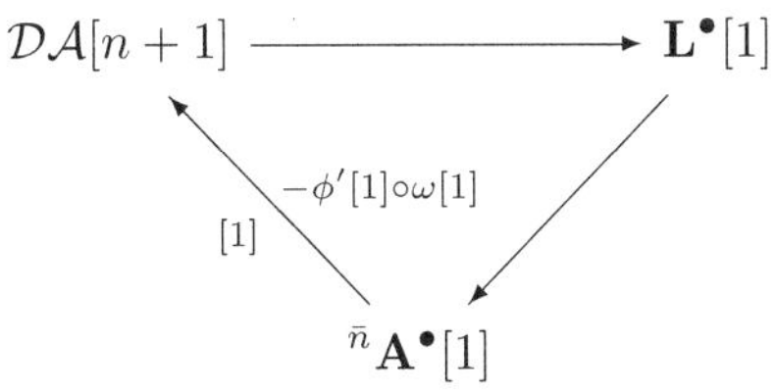

Dualizing and shifting the $\boxplus$-triangle yields

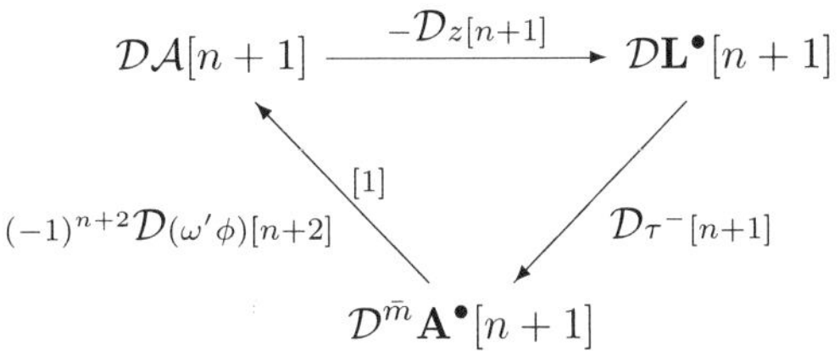

Using lemma 2.3, we have the isomorphism $\delta : \mathcal{D}\mathcal{O}(\mathbf{A}^\bullet)[n+1] \xrightarrow{\cong} \mathcal{O}(\mathbf{A}^\bullet)$ such that

$$\begin{array}{ccc} \mathcal{D}^{\bar m}\mathbf{A}^\bullet[n+1] & \xrightarrow{\mathcal{D}\omega'[n+2]} & \mathcal{D}\mathcal{O}(\mathbf{A}^\bullet)[n+2] \\ {\scriptstyle d'[1]}\downarrow\cong & & \cong\downarrow{\scriptstyle \delta[1]} \\ {}^{\bar n}\mathbf{A}^\bullet[1] & \xrightarrow{\omega[1]} & \mathcal{O}(\mathbf{A}^\bullet)[1] \end{array}$$

commutes. As ϕ is a Lagrangian structure, we have $\delta \in \{d_\gamma\}$, whence

$$\begin{array}{ccc} \mathcal{D}\mathcal{O}(\mathbf{A}^\bullet)[n+2] & \xrightarrow{(-1)^{n+2}\mathcal{D}\phi[n+2]} & \mathcal{D}\mathcal{A}[n+2] \\ {\scriptstyle \delta[1]}\downarrow\cong & & \Vert \\ \mathcal{O}(\mathbf{A}^\bullet)[1] & \xrightarrow{-\phi'[1]} & \mathcal{D}\mathcal{A}[n+2] \end{array}$$

commutes, compare (4). Putting the above two squares together, we obtain that

$$\begin{array}{ccc} \mathcal{D}^{\bar m}\mathbf{A}^\bullet[n+1] & \xrightarrow{(-1)^{n+2}\mathcal{D}(\omega'\phi)[n+2]} & \mathcal{D}\mathcal{A}[n+2] \\ {\scriptstyle d'[1]}\downarrow\cong & & \Vert \\ {}^{\bar n}\mathbf{A}^\bullet[1] & \xrightarrow{-\phi'[1]\circ\omega[1]} & \mathcal{D}\mathcal{A}[n+2] \end{array}$$

commutes. This square can be embedded in an isomorphism of triangles, inducing d:

$$\begin{array}{ccc} \mathcal{D}\mathcal{A}[n+1] & = & \mathcal{D}\mathcal{A}[n+1] \\ {\scriptstyle -\mathcal{D}z[n+1]}\downarrow & & \downarrow \\ \mathcal{D}\mathbf{L}^\bullet[n+1] & \underset{\cong}{\xrightarrow{d[1]}} & \mathbf{L}^\bullet[1] \\ {\scriptstyle \mathcal{D}\tau^-[n+1]}\downarrow & & \downarrow \\ \mathcal{D}^{\bar m}\mathbf{A}^\bullet[n+1] & \underset{\cong}{\xrightarrow{d'[1]}} & {}^{\bar n}\mathbf{A}^\bullet[1] \\ {\scriptstyle (-1)^{n+2}\mathcal{D}(\omega'\phi)[n+2]}\downarrow & & \downarrow{\scriptstyle -\phi'[1]\circ\omega[1]} \\ \mathcal{D}\mathcal{A}[n+2] & = & \mathcal{D}\mathcal{A}[n+2] \end{array}$$

We have $\mathbf{H}^i(\mathbf{A}^\bullet \boxplus \mathcal{A}|_{U_{l+1}}) = 0$ for $i > \bar n(l) - n$, $l \geq 2$ and $\mathbf{H}^i(j_x^! {}^{\bar n}\mathbf{A}^\bullet) = 0$ for $i \leq \bar n(l) - l$, $x \in X_{n-l} - X_{n-l-1}$ (see proof of theorem 2.6). Thus, restriction is injective ([**CS91**], 1.3)

$$\mathrm{Hom}_{D^b(U_{k+1})}(\mathbf{L}^\bullet, {}^{\bar n}\mathbf{A}^\bullet) \hookrightarrow \mathrm{Hom}_{D^b(U_2)}(\mathbf{L}^\bullet|_{U_2}, {}^{\bar n}\mathbf{A}^\bullet|_{U_2}) \qquad (19)$$

By the commutativity relations in an octahedron, the map $\kappa : \mathbf{L}^\bullet \to {}^{\bar{n}}\mathbf{A}^\bullet$ in (18) satisfies

$$\kappa\tau^- = c$$

and so on U_2,

$$\kappa|_{U_2} \circ 1_{\mathbf{A}^\bullet|U_2} = \kappa|_{U_2} \circ \tau^-|_{U_2} = c|_{U_2} = 1_{\mathbf{A}^\bullet|U_2}$$

(using canonical identifications), so that

$$\kappa|_{U_2} = 1$$

But also $\tau^+|_{U_2} = 1$, and thus the injectivity statement (19) implies $\kappa = \tau^+$. Again using [**CS91**] (1.3), we have an injection

$$\text{(20)} \qquad \mathrm{Hom}_{D^b(U_{k+1})}(\mathcal{D}\mathbf{L}^\bullet[n], \mathbf{L}^\bullet) \hookrightarrow \mathrm{Hom}_{D^b(U_2)}(\mathcal{D}\mathbf{L}^\bullet|_{U_2}[n], \mathbf{L}^\bullet|_{U_2})$$

Let $d_A : \mathcal{D}\mathbf{A}^\bullet[n] \xrightarrow{\cong} \mathbf{A}^\bullet$, $\mathcal{D}d_A[n] = d_A$, denote the self-duality isomorphism associated to $\mathbf{A}^\bullet$ by (**SD4**).
Since $d'|_{U_k} = d_A$ ([**GM83**], section 9), we have

$$d'|_{U_2} = d_A|_{U_2} = \mathcal{D}d_A|_{U_2}[n] = \mathcal{D}d'|_{U_2}[n].$$

The middle commutative square in the above isomorphism of triangles gives the relation

$$\tau^+ d = d'\mathcal{D}\tau^-[n]$$

and hence

$$\begin{aligned} d|_{U_2} &= 1 \circ d|_{U_2} = \tau^+|_{U_2} \circ d|_{U_2} = d'|_{U_2} \circ \mathcal{D}\tau^-|_{U_2}[n] \\ &= d'|_{U_2} \circ \mathcal{D}1[n] = d'|_{U_2}. \end{aligned}$$

Therefore,

$$\mathcal{D}d|_{U_2}[n] = \mathcal{D}d'|_{U_2}[n] = d'|_{U_2} = d|_{U_2}$$

and by the injectivity of (20), $\mathcal{D}d[n] = d$.

□

6. A Postnikov system

In this section we obtain our main results concerning the algebraic structure of the category $SD(X)$ of self-dual sheaves compatible with intersection homology theory on non-Witt spaces. The functors Λ and $\boxplus$ are being used in constructing an equivalence of categories between $SD(X)$ and a Postnikov system whose fibers are categories of Lagrangian structures along the various pure singular strata of odd codimension.

Consider the functor

$$(i_k^*, \Lambda) : SD(U_{k+1}) \longrightarrow SD(U_k) \rtimes \mathrm{Lag}(U_{k+1} - U_k).$$

THEOREM 2.8. *There is an isomorphism of functors*

$$\boxplus(i_k^*, \Lambda) \cong 1_{SD(U_{k+1})}$$

PROOF. Given $\mathbf{E}^\bullet \in SD(U_{k+1})$, we have to construct a natural isomorphism

$$T(\mathbf{E}^\bullet) : \mathbf{E}^\bullet \xrightarrow{\cong} i_k^* \mathbf{E}^\bullet \boxplus \Lambda(\mathbf{E}^\bullet)$$

Write $\Lambda(\mathbf{E}^\bullet) = (\mathcal{L} \xrightarrow{\phi} \mathcal{O}(\mathbf{E}^\bullet|))$. Then by theorem 2.3 we have the distinguished triangle

$$\begin{array}{ccc} {}^{\bar m}\mathbf{E}^\bullet| & \xrightarrow{\quad a \quad} & \mathbf{E}^\bullet \\ & {\scriptstyle [1]}\nwarrow^{\alpha'} \quad \swarrow_{\alpha} & \\ & \mathcal{L} & \end{array}$$

Recall that from the octahedron used to define the Lagrangian structure triangle (9), α' satisfies

$$\alpha' = \omega' \phi,$$

where ω' is the morphism raising degrees in the lifting obstruction triangle, theorem 2.1, associated to the restriction $\mathbf{E}^\bullet|$.
On the other hand, by theorem 2.6, we have the $\boxplus$-triangle

$$\begin{array}{ccc} \mathcal{L} & \xrightarrow{\quad \omega'\phi \quad} & {}^{\bar m}\mathbf{E}^\bullet|[1] \\ & {\scriptstyle [1]}\nwarrow^{-z[1]} \quad \swarrow_{-\tau^-[1]} & \\ & (i_k^*\mathbf{E}^\bullet \boxplus \Lambda(\mathbf{E}^\bullet))[1] & \end{array}$$

Embedding the commutative square

$$\begin{array}{ccc} \mathcal{L} & \xrightarrow{\alpha'} & {}^{\bar m}\mathbf{E}^\bullet|[1] \\ \| & & \| \\ \mathcal{L} & \xrightarrow{\omega'\phi} & {}^{\bar m}\mathbf{E}^\bullet|[1] \end{array}$$

into an isomorphism of triangles produces $T(\mathbf{E}^\bullet)$:

(21)
$$\begin{array}{ccccccc} \mathcal{L} & \xrightarrow{\alpha'} & {}^{\bar m}\mathbf{E}^\bullet|[1] & \xrightarrow{-a[1]} & \mathbf{E}^\bullet[1] & \xrightarrow{-\alpha[1]} & \mathcal{L}[1] \\ \| & & \| & & {\scriptstyle T(\mathbf{E}^\bullet)}\downarrow\cong & & \| \\ \mathcal{L} & \xrightarrow{\omega'\phi} & {}^{\bar m}\mathbf{E}^\bullet|[1] & \xrightarrow{-\tau^-[1]} & (i_k^*\mathbf{E}^\bullet \boxplus \Lambda(\mathbf{E}^\bullet))[1] & \xrightarrow{-z[1]} & \mathcal{L}[1] \end{array}$$

Given $f : \mathbf{E}^\bullet \to \mathbf{F}^\bullet$, with $\mathbf{E}^\bullet, \mathbf{F}^\bullet \in SD(U_{k+1})$ one has to verify that

$$\begin{array}{ccc} \mathbf{E}^\bullet & \xrightarrow{T(\mathbf{E}^\bullet)} & i_k^*\mathbf{E}^\bullet \boxplus \Lambda(\mathbf{E}^\bullet) \\ {\scriptstyle f}\big\downarrow & & \big\downarrow{\scriptstyle f| \boxplus \Lambda(f)} \\ \mathbf{F}^\bullet & \xrightarrow{T(\mathbf{F}^\bullet)} & i_k^*\mathbf{F}^\bullet \boxplus \Lambda(\mathbf{F}^\bullet) \end{array} \tag{22}$$

commutes. Using canonical identifications (see proposition 2.6), we have $\tau^-|_{U_2} = 1_{\mathbf{F}^\bullet|_{U_2}}$, $a|_{U_2} = 1_{\mathbf{F}^\bullet|_{U_2}}$, and hence, by the middle square of (21)

$$\begin{aligned} T(\mathbf{E}^\bullet)|_{U_2} &= 1_{\mathbf{E}^\bullet|U_2}, \\ T(\mathbf{F}^\bullet)|_{U_2} &= 1_{\mathbf{F}^\bullet|U_2}. \end{aligned}$$

By lemma 2.2, restriction to U_2 is an injection

$$\mathrm{Hom}_{D^b(U_{k+1})}(\mathbf{E}^\bullet, i_k^*\mathbf{F}^\bullet \boxplus \Lambda(\mathbf{F}^\bullet)) \hookrightarrow \mathrm{Hom}_{D^b(U_2)}(\mathbf{E}^\bullet|_{U_2}, (i_k^*\mathbf{F}^\bullet \boxplus \Lambda(\mathbf{F}^\bullet))|_{U_2})$$

and it suffices to verify the commutativity of (22) over U_2. But this amounts to seeing

$$f|_{U_2} = (f|_{U_k} \boxplus \Lambda(f))|_{U_2}$$

which is true by proposition 2.6. □

THEOREM 2.9. *There is an isomorphism of functors*

$$(i_k^*, \Lambda)\boxplus \cong 1_{SD(U_k) \rtimes \mathrm{Lag}(U_{k+1} - U_k)}$$

PROOF. Given $\mathbf{A}^\bullet \in SD(U_k)$ and $(\mathcal{A} \xrightarrow{\phi} \mathcal{O}(\mathbf{A}^\bullet)) \in \mathrm{Lag}(U_{k+1} - U_k)$, we have to construct a natural isomorphism

$$T(\mathbf{A}^\bullet, \mathcal{A}) : (\mathbf{A}^\bullet, \mathcal{A}) \xrightarrow{\cong} (i_k^*, \Lambda)(\mathbf{A}^\bullet \boxplus \mathcal{A}),$$

i.e. we need two components

$$\begin{aligned} T_1 &: \mathbf{A}^\bullet \xrightarrow{\cong} i_k^*(\mathbf{A}^\bullet \boxplus \mathcal{A}), \text{and} \\ T_2 &: (\mathcal{A} \xrightarrow{\phi} \mathcal{O}(\mathbf{A}^\bullet)) \xrightarrow{\cong} \Lambda(\mathbf{A}^\bullet \boxplus \mathcal{A}). \end{aligned}$$

Take T_1 to be $T_1 = \rho$, where ρ is the canonical identification from proposition 2.6. Next, we discuss how to obtain T_2.
First, we remark that

$$\begin{array}{ccc} {}^{\bar{m}}\mathbf{A}^\bullet & \xrightarrow{\tau^-} & \mathbf{A}^\bullet \boxplus \mathcal{A} \\ {\scriptstyle {}^{\bar{m}}\rho}\big\downarrow{\scriptstyle \cong} & & \Big\| \\ {}^{\bar{m}}(\mathbf{A}^\bullet \boxplus \mathcal{A})| & \xrightarrow{a} & \mathbf{A}^\bullet \boxplus \mathcal{A} \end{array} \tag{23}$$

commutes, as follows from ${}^{\bar{m}}\rho|_{U_2} = 1_{\mathbf{A}^\bullet|U_2}$, $a|_{U_2} = 1_{\mathbf{A}^\bullet|U_2}$, $\tau^-|_{U_2} = 1_{\mathbf{A}^\bullet|U_2}$ and

$$\mathrm{Hom}_{D^b(U_{k+1})}({}^{\bar{m}}\mathbf{A}^\bullet, \mathbf{A}^\bullet \boxplus \mathcal{A}) \hookrightarrow \mathrm{Hom}_{D^b(U_2)}(\mathbf{A}^\bullet|_{U_2}, \mathbf{A}^\bullet|_{U_2}),$$

observing $\mathbf{H}^i({}^{\bar m}\mathbf{A}^\bullet|_{U_{l+1}}) = 0$ for $i > \bar n(l) - n$, $l \geq 2$ and $\mathbf{H}^i(j_x^!(\mathbf{A}^\bullet \boxplus \mathcal{A})) = 0$ for $i \leq \bar n(l) - l$, $x \in X_{n-l} - X_{n-l-1}$ (lemma 2.1) and applying [**CS91**], (1.3).
Since $\mathbf{H}^i(j_x^! {}^{\bar n}(\mathbf{A}^\bullet \boxplus \mathcal{A})|) = 0$ for $i \leq \bar n(l) - l$, an analogous argument implies that

$$\begin{array}{ccc} {}^{\bar n}\mathbf{A}^\bullet & \xrightarrow{{}^{\bar n}\rho} & {}^{\bar n}(\mathbf{A}^\bullet \boxplus \mathcal{A})| \\ {\scriptstyle \tau^+}\uparrow & & \| \\ \mathbf{A}^\bullet \boxplus \mathcal{A} & \xrightarrow{b} & {}^{\bar n}(\mathbf{A}^\bullet \boxplus \mathcal{A})| \end{array} \tag{24}$$

commutes.
We write $\Lambda(\mathbf{A}^\bullet \boxplus \mathcal{A}) = (\mathcal{L} \xrightarrow{\psi} \mathcal{O}(\mathbf{A}^\bullet \boxplus \mathcal{A}|))$. The square (23) can be embedded in an isomorphism of triangles

$$\begin{array}{ccccccc} {}^{\bar m}\mathbf{A}^\bullet & \xrightarrow{\tau^-} & \mathbf{A}^\bullet \boxplus \mathcal{A} & \xrightarrow{z} & \mathcal{A} & \xrightarrow{\omega'\phi} & {}^{\bar m}\mathbf{A}^\bullet[1] \\ {\scriptstyle {}^{\bar m}\rho}\downarrow{\scriptstyle \cong} & & \| & & {\scriptstyle l}\downarrow{\scriptstyle \cong} & & \downarrow{\scriptstyle \cong} \\ {}^{\bar m}(\mathbf{A}^\bullet \boxplus \mathcal{A})| & \xrightarrow{a} & \mathbf{A}^\bullet \boxplus \mathcal{A} & \xrightarrow{\alpha} & \mathcal{L} & \xrightarrow{\alpha'} & {}^{\bar m}(\mathbf{A}^\bullet \boxplus \mathcal{A})|[1] \end{array}$$

inducing an isomorphism $l : \mathcal{A} \to \mathcal{L}$. (Here, the upper row is the $\boxplus$-triangle, theorem 2.6, and the lower row is given by theorem 2.3.)
We have to check that

$$\begin{array}{ccc} \mathcal{A} & \xrightarrow{\phi} & \mathcal{O}(\mathbf{A}^\bullet) \\ {\scriptstyle l}\downarrow{\scriptstyle \cong} & & {\scriptstyle \cong}\downarrow{\scriptstyle \mathcal{O}(\rho)} \\ \mathcal{L} & \xrightarrow{\psi} & \mathcal{O}(\mathbf{A}^\bullet \boxplus \mathcal{A}|) \end{array}$$

commutes. We have commutative squares

$$\begin{array}{ccc} \mathbf{A}^\bullet \boxplus \mathcal{A} & \xrightarrow{b} & {}^{\bar n}(\mathbf{A}^\bullet \boxplus \mathcal{A})| \\ {\scriptstyle \alpha}\downarrow & & \downarrow{\scriptstyle \omega} \\ \mathcal{L} & \xrightarrow{\psi} & \mathcal{O}(\mathbf{A}^\bullet \boxplus \mathcal{A}|) \end{array}$$

(see remark 2.5) and (15)

$$\begin{array}{ccc} \mathbf{A}^\bullet \boxplus \mathcal{A} & \xrightarrow{\tau^+} & {}^{\bar n}\mathbf{A}^\bullet \\ {\scriptstyle z}\downarrow & & \downarrow{\scriptstyle \omega} \\ \mathcal{A} & \xrightarrow{\phi} & \mathcal{O}(\mathbf{A}^\bullet) \end{array}$$

The square

$$\begin{array}{ccc}
{}^{\bar n}\mathbf{A}^\bullet & \xrightarrow{{}^{\bar n}\rho} & {}^{\bar n}(\mathbf{A}^\bullet \boxplus \mathcal{A})| \\
\Big\downarrow{\scriptstyle \omega} & & \Big\downarrow{\scriptstyle \omega} \\
\mathcal{O}(\mathbf{A}^\bullet) & \xrightarrow{\mathcal{O}(\rho)} & \mathcal{O}(\mathbf{A}^\bullet \boxplus \mathcal{A}|)
\end{array}$$

commutes for trivial reasons. Putting these squares and (24) together, we obtain the cube

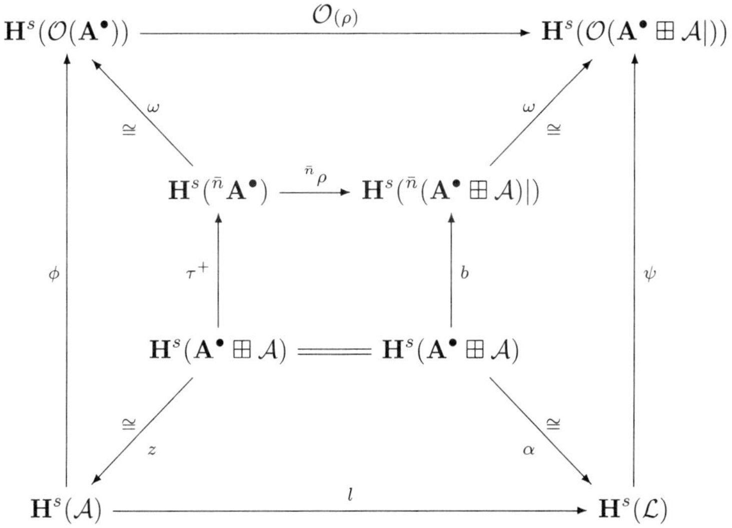

All faces with at least one edge in the inside commute. Hence, using that $\mathbf{H}^s(\omega)$ and $\mathbf{H}^s(z)$ are isomorphisms, the outside square commutes. Then

$$\psi l = \mathcal{O}(\rho)\phi$$

follows from

$$\mathrm{Hom}_{D^b(U_{k+1})}(\mathcal{A}, \mathcal{O}(\mathbf{A}^\bullet \boxplus \mathcal{A}|)) \cong \mathrm{Hom}(\mathbf{H}^s(\mathcal{A}), \mathbf{H}^s(\mathcal{O}(\mathbf{A}^\bullet \boxplus \mathcal{A}|))).$$

We define

$$T_2 : (\mathcal{A} \xrightarrow{\phi} \mathcal{O}(\mathbf{A}^\bullet)) \xrightarrow{\cong} \Lambda(\mathbf{A}^\bullet \boxplus \mathcal{A})$$

to be the square

$$\begin{array}{ccc}
\mathcal{A} & \xrightarrow{\phi} & \mathcal{O}(\mathbf{A}^\bullet) \\
{\scriptstyle l}\Big\downarrow{\scriptstyle \cong} & & {\scriptstyle \cong}\Big\downarrow{\scriptstyle \mathcal{O}(\rho)} \\
\mathcal{L} & \xrightarrow{\psi} & \mathcal{O}(\mathbf{A}^\bullet \boxplus \mathcal{A}|)
\end{array}$$

(The embedding of a commutative square in a morphism of triangles is not in general unique. However, by remark 2.3, the l in the present case is unique).

It remains to be shown that T_2 is a natural transformation. Suppose we are given $\mathbf{A}^\bullet, \mathbf{B}^\bullet \in SD(U_k)$, Lagrangian structures $(\mathcal{A} \xrightarrow{\phi_A} \mathcal{O}(\mathbf{A}^\bullet))$, $(\mathcal{B} \xrightarrow{\phi_B} \mathcal{O}(\mathbf{B}^\bullet)) \in \mathrm{Lag}(U_{k+1} - U_k)$, and a morphism in $SD(U_k) \rtimes \mathrm{Lag}(U_{k+1} - U_k)$:

$$(f, t) : (\mathbf{A}^\bullet, \mathcal{A}) \longrightarrow (\mathbf{B}^\bullet, \mathcal{B})$$

We have to verify that

$$\begin{array}{ccc} (\mathcal{A} \xrightarrow{\phi_A} \mathcal{O}(\mathbf{A}^\bullet)) & \underset{\cong}{\xrightarrow{T_2}} & \Lambda(\mathbf{A}^\bullet \boxplus \mathcal{A}) \\ \Big\downarrow t & & \Big\downarrow \Lambda(f \boxplus t) \\ (\mathcal{B} \xrightarrow{\phi_B} \mathcal{O}(\mathbf{B}^\bullet)) & \underset{\cong}{\xrightarrow{T_2}} & \Lambda(\mathbf{B}^\bullet \boxplus \mathcal{B}) \end{array}$$

commutes. More precisely, if

$$\Lambda(\mathbf{A}^\bullet \boxplus \mathcal{A}) = (\mathcal{L}_A \xrightarrow{\psi_A} \mathcal{O}(\mathbf{A}^\bullet \boxplus \mathcal{A}|)),$$
$$\Lambda(\mathbf{B}^\bullet \boxplus \mathcal{B}) = (\mathcal{L}_B \xrightarrow{\psi_B} \mathcal{O}(\mathbf{B}^\bullet \boxplus \mathcal{B}|)),$$

and

$$T_2(\mathbf{A}^\bullet, \mathcal{A}) = \begin{array}{ccc} \mathcal{A} & \xrightarrow{\phi_A} & \mathcal{O}(\mathbf{A}^\bullet) \\ l_A \downarrow\cong & & \cong \downarrow \mathcal{O}(\rho) \\ \mathcal{L}_A & \xrightarrow{\psi_A} & \mathcal{O}(\mathbf{A}^\bullet \boxplus \mathcal{A}|) \end{array}$$

$$T_2(\mathbf{B}^\bullet, \mathcal{B}) = \begin{array}{ccc} \mathcal{B} & \xrightarrow{\phi_B} & \mathcal{O}(\mathbf{B}^\bullet) \\ l_B \downarrow\cong & & \cong \downarrow \mathcal{O}(\rho) \\ \mathcal{L}_B & \xrightarrow{\psi_B} & \mathcal{O}(\mathbf{B}^\bullet \boxplus \mathcal{B}|) \end{array}$$

we need to see that the squares

(25) $$\begin{array}{ccc} \mathcal{O}(\mathbf{A}^\bullet) & \xrightarrow{\mathcal{O}(f)} & \mathcal{O}(\mathbf{B}^\bullet) \\ \mathcal{O}(\rho) \Big\downarrow \cong & & \cong \Big\downarrow \mathcal{O}(\rho) \\ \mathcal{O}(\mathbf{A}^\bullet \boxplus \mathcal{A}|) & \xrightarrow{\mathcal{O}(f \boxplus t|)} & \mathcal{O}(\mathbf{B}^\bullet \boxplus \mathcal{B}|) \end{array}$$

and

(26) $$\begin{array}{ccc} \mathcal{A} & \xrightarrow{l_A} & \mathcal{L}_A \\ t \Big\downarrow & & \Big\downarrow \lambda \\ \mathcal{B} & \xrightarrow{l_B} & \mathcal{L}_B \end{array}$$

commute, where $\lambda = \Lambda(f \boxplus t)$.
By proposition 2.6 we have a commutative diagram

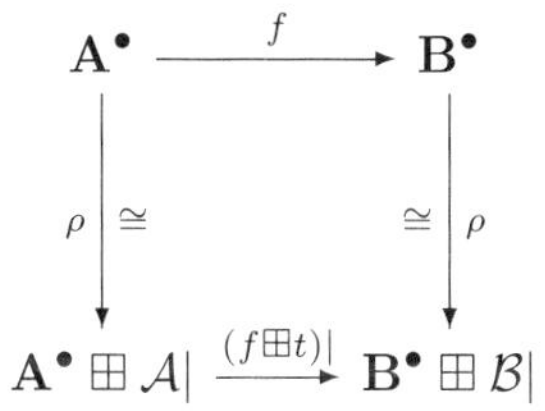

Applying the functor $\mathcal{O}$, we see that (25) commutes. We proceed to (26). Consider the following cube:

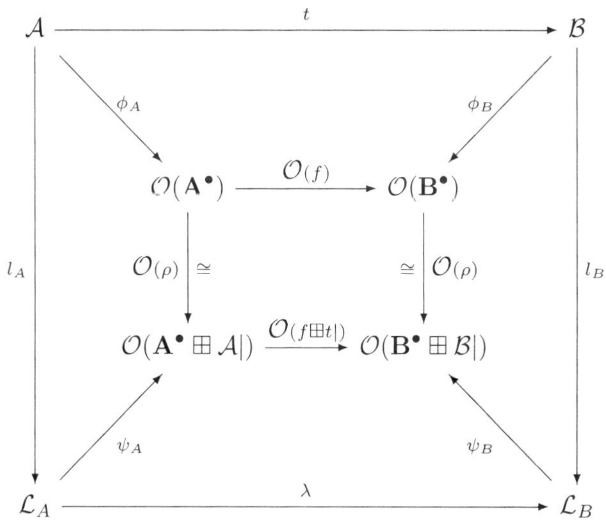

All faces having at least one inside edge commute. The following calculation establishes that the outside square commutes as well.

On $\mathbf{H}^s(\cdot)$, ϕ_B is injective, hence invertible on its image and we have

$$\mathbf{H}^s(t) = \mathbf{H}^s(\phi_B)|^{-1}\mathbf{H}^s(\mathcal{O}(f))\mathbf{H}^s(\phi_A)$$

A similar remark applies to ψ_B. We compute

$$\begin{aligned}
\mathbf{H}^s(l_B)\mathbf{H}^s(t) &= \mathbf{H}^s(l_B)\mathbf{H}^s(\phi_B)|^{-1}\mathbf{H}^s(\mathcal{O}(f))\mathbf{H}^s(\phi_A) \\
&= \mathbf{H}^s(\psi_B)|^{-1}\mathbf{H}^s(\mathcal{O}(\rho))\mathbf{H}^s(\mathcal{O}(f))\mathbf{H}^s(\phi_A) \\
&= \mathbf{H}^s(\psi_B)|^{-1}\mathbf{H}^s(\mathcal{O}(f \boxplus t|))\mathbf{H}^s(\mathcal{O}(\rho))\mathbf{H}^s(\phi_A) \\
&= \mathbf{H}^s(\psi_B)|^{-1}\mathbf{H}^s(\mathcal{O}(f \boxplus t|))\mathbf{H}^s(\psi_A)\mathbf{H}^s(l_A) \\
&= \mathbf{H}^s(\lambda)\mathbf{H}^s(l_A).
\end{aligned}$$

This implies $l_B t = \lambda l_A$, as

$$\mathrm{Hom}_{D^b(U_{k+1})}(\mathcal{A}, \mathcal{L}_B) \cong \mathrm{Hom}(\mathbf{H}^s(\mathcal{A}), \mathbf{H}^s(\mathcal{L}_B)).$$

□

Let $\mathrm{Const}(U_2)$ denote the category whose single object is the constant sheaf $\mathbb{R}_{U_2}$ on U_2 and whose morphisms are sheaf maps

$$\mathbb{R}_{U_2} \to \mathbb{R}_{U_2}.$$

We are now in a position to state our main result, a Postnikov-type decomposition of the category $SD(X)$:

THEOREM 2.10. *Let $n = \dim X$ be even. There is an equivalence of categories*

$$SD(X) \simeq \mathrm{Lag}(U_n - U_{n-1}) \rtimes \mathrm{Lag}(U_{n-2} - U_{n-3}) \rtimes \ldots \rtimes \mathrm{Lag}(U_4 - U_3) \rtimes \mathrm{Const}(U_2).$$

PROOF. The functor

$$p : SD(U_2) \longrightarrow \mathrm{Const}(U_2)$$

defined by $p(\mathbf{S}^\bullet) = \mathbb{R}_{U_2}$, for $\mathbf{S}^\bullet \in SD(U_2)$ and $p(f) = \nu_2^{-1} f \nu_1$ for $f : \mathbf{S}_1^\bullet \to \mathbf{S}_2^\bullet$, $\mathbf{S}_1^\bullet, \mathbf{S}_2^\bullet \in SD(U_2)$ (ν_i is the normalization associated to $\mathbf{S}_i^\bullet, i = 1, 2$) is a pseudoinverse for the natural inclusion

$$\mathrm{Const}(U_2) \hookrightarrow SD(U_2).$$

Hence, we have an equivalence of categories

$$SD(U_2) \simeq \mathrm{Const}(U_2).$$

Suppose $k \geq 2$ is even. In that case, $\bar{m}(k) = \bar{n}(k)$ and the functor

$$\tau_{\leq \bar{m}(k) - n} R i_{k*} : SD(U_k) \longrightarrow D^b(U_{k+1})$$

takes values in $SD(U_{k+1})$, as for $\mathbf{A}^\bullet \in SD(U_k)$,

$$\mathcal{D}({}^{\bar{m}}\mathbf{A}^\bullet)[n] \cong {}^{\bar{n}}\mathbf{A}^\bullet = {}^{\bar{m}}\mathbf{A}^\bullet$$

by [**GM83**], section 9. Its pseudoinverse is restriction

$$i_k^* : SD(U_{k+1}) \longrightarrow SD(U_k)$$

Thus, we have an equivalence of categories

$$SD(U_{k+1}) \simeq SD(U_k).$$

If $k \geq 2$ is odd, then theorems 2.8 and 2.9 give an equivalence of categories

$$SD(U_{k+1}) \simeq SD(U_k) \rtimes \mathrm{Lag}(U_{k+1} - U_k).$$

The statement follows by successively using these equivalences. □

This theorem gives precise information as to which data is needed to extend generalized Poincaré duality and intersection homology to spaces that do not satisfy the Witt space condition. For odd-dimensional X, see the remarks in section 3 of the introduction, chapter 1.

CHAPTER 3

Ordered Resolutions

1. The Purpose of the Construction

We will construct a class of pseudomanifolds which may or may not be Witt, depending on the input parameters. This class of spaces is distinguished by the property that the very nature of their geometry allows for the construction of self-dual (complexes of) sheaves which interpolate between their lower middle perversity and upper middle perversity intersection homology sheaves. The idea is to achieve a geometry where each link of each nonsingular stratum is the boundary of some pseudomanifold on which, by an inductive hypothesis, a self-dual sheaf has already been constructed. This will give rise to a preferred Lagrangian subspace in the hypercohomology of the link with coefficients in the already available sheaf.

Our production of this class of spaces can be outlined as follows: Start with a closed oriented manifold M, embed a closed manifold E, which is the total space of a fiber bundle over a manifold base B. Define the space $X^{(1)}$ as the fiberwise quotient of M with respect to the fibration. $X^{(1)}$ is a pseudomanifold with one singular stratum B. Now assume that $X^{(m)}$ has been constructed. Let $X_k^{(m)}$ be its lowest dimensional singular stratum. Embed a closed manifold $E_m \hookrightarrow X_k^{(m)}$, which is the total space of a fiber bundle over a manifold base $B^{b(m)}$, where $b(m)$ denotes the dimension. Define $X^{(m+1)}$ as the fiberwise quotient of $X^{(m)}$ with respect to the fibration. $X^{(m+1)}$ is a pseudomanifold with $m+1$ singular strata. On every space in the thus obtained (finite) sequence $\{X^{(m)}\}$, a self-dual sheaf can be constructed, starting with the manifold $X^{(0)} = M$.

The category that we choose to work in is the PL-category. The motivation for this choice is to be able to use block bundle theory as developed by Rourke and Sanderson [**RS68**] and the theory of PL-varieties and cone block bundles as developed by Stone [**Sto72**], as these theories enable us to obtain rather precise information about the neighborhoods and links of strata. So the spaces that we shall construct will actually be PL-varieties.

2. Definitions

If L is a polyhedron, cL denotes the cone on L, c being the cone point.
Let X be a polyhedron. A filtration $X = X_n \supset X_{n-1} \supset X_{n-2} \supset \ldots \supset X_0$ by subpolyhedra is called a *(PL-) variety* (see [**Sto72**] or [**McC75**]) if for each point x in X, say $x \in X_i - X_{i-1}$, there is a compact polyhedron L (the link of the open stratum $X_i - X_{i-1}$) such that a neighborhood N of x in X is isomorphic to $D^i \times cL$, with $X_i \cap N$ corresponding to $D^i \times \{c\}$ and, for each $j > i$, $X_j \cap N$ corresponding to

$D^i \times cL_j$ for some subpolyhedron $L_j \subset L$. It follows that each nonempty $X_i - X_{i-1}$ is an i-manifold.

By a *block fibration* (with fiber F) we mean a map $\pi : E(\xi) \to B$, where $E(\xi)$ is the total space of an F-block bundle ξ/K over the cell complex K, $|K| = B$, such that for every $\sigma \in K$,

1. $\pi^{-1}(|\sigma|) = |\xi(\sigma)|$, where $\xi(\sigma)$ is the block over σ, and
2. $\xi(\sigma)$ has a structure $\phi : |\xi(\sigma)| \xrightarrow{\cong} D^k \times F$ ($k = \dim \sigma$) which commutes with the projection:

$$\begin{array}{ccc} |\xi(\sigma)| & \xrightarrow{\phi} & D^k \times F \\ \Big\downarrow{\scriptstyle \pi|} & & \Big\downarrow{\scriptstyle \text{proj}} \\ |\sigma| & \xrightarrow{\phi|} & D^k \end{array}$$

LEMMA 3.1. *Let $N^n \hookrightarrow M^m$ be a locally flat embedding of one PL-manifold into another one. If there exists a normal block bundle ξ/K, $|K| = N$, for $N \hookrightarrow M$ that can be equipped with a projection $E(\xi) \xrightarrow{\pi} N$ making ξ into a block fibration, then every normal block bundle η/L, $|L| = N$, for $N \hookrightarrow M$ can be equipped with such a projection.*

PROOF. Let J be a common subdivision of K and L. Let η'/J be a subdivision of η and, using the projection, define the subdivision ξ'/J as $\xi' = \{\pi^{-1}(|\sigma|) : \sigma \in J\}$. By Rourke and Sanderson's uniqueness theorem [**RS68**], there exists a block bundle isomorphism $h : E(\eta') \xrightarrow{\cong} E(\xi')$. Define $p = \pi h$. We have to show that $p : E(\eta') \to N$ makes η' (and therefore η) into a block fibration. Let $\eta'(\sigma)$ be a block of η', $\sigma \in J$. $\xi'(\sigma)$ has a structure

$$\begin{array}{ccc} |\xi'(\sigma)| & \xrightarrow{\phi} & D^k \times D^{m-n} \\ \Big\downarrow{\scriptstyle \pi|} & & \Big\downarrow{\scriptstyle \text{proj}} \\ |\sigma| & \xrightarrow{\phi|} & D^k \end{array}$$

where $k = \dim \sigma$. A structure for $\eta'(\sigma)$ is given by

$$\begin{array}{ccccc} |\eta'(\sigma)| & \xrightarrow{h|} & |\xi'(\sigma)| & \xrightarrow{\phi} & D^k \times D^{m-n} \\ \Big\downarrow{\scriptstyle h|} & & \Big\| & & \Big\| \\ |\xi'(\sigma)| & = & |\xi'(\sigma)| & \xrightarrow{\phi} & D^k \times D^{m-n} \\ \Big\downarrow{\scriptstyle \pi|} & & \Big\downarrow{\scriptstyle \pi|} & & \Big\downarrow{\scriptstyle \text{proj}} \\ |\sigma| & = & |\sigma| & \xrightarrow{\phi|} & D^k \end{array}$$

□

Let N and M be PL-manifolds. Call $f : N \hookrightarrow M$ a *good embedding* if it is locally flat and some normal block bundle for f is a block fibration.

Remarks:

1. If f is locally flat, then by Rourke and Sanderson's existence theorem [**RS68**], f has a normal block bundle.

2. If f is a good embedding, then the lemma shows that every normal block bundle is a block fibration.

We shall also use the following notation: If K is a simplicial complex, K' a first derived, and $\sigma \in K$, let $\sigma^*(K')$ be the dual block of σ with respect to K'.

A variety analog of lemma 3.1 is

LEMMA 3.2. *Let $X^n = X_n \supset X_{n-1} \supset \ldots \supset X_k$ be a PL-variety. If there exists a normal cone block bundle ξ/K, $|K| = X_k$ that can be equipped with a projection $E(\xi) \xrightarrow{\pi} X_k$ making ξ into a block fibration, then every normal cone block bundle over X_k can be equipped with such a projection.*

PROOF. Same as for lemma 3.1, only use Stone's uniqueness theorem [**Sto72**] instead to obtain the bundle isomorphism. □

3. The PL Construction

We now proceed to the details of the construction. The input data consist of the following: Let $X^n = X_n \supset X_{n-1} \supset \ldots \supset X_k$ be a closed PL-variety such that a blocking of a regular neighborhood of X_k in X is a block fibration. Note that the bottom stratum X_k is a manifold. Let $\pi : E^l \to B^b$ be a block fibration with fiber F_0, with E, B and F_0 closed PL-manifolds and let $E \hookrightarrow X_k$ be a good embedding. We assume E, X_k to be connected. By making suitable modifications and working over connected components, it can be seen that this assumption is not essential. We identify E with its image under the embedding.

We shall describe how to associate with this data a new PL-variety Y which can be thought of as a homeomorph of the fiberwise quotient $X/\sim$, where $x \sim y$ iff $x, y \in E$ and $\pi(x) = \pi(y)$. The variety data will enable us to give a detailed account of the geometry of Y (e.g. what is the link of the newly generated bottom stratum in Y ?)

Step 1. We will triangulate X so that the regular neighborhoods of E in X_k, of X_k in X and of E in X can be simultaneously described as derived neighborhoods: Triangulate X by a simplicial complex K such that X_k is triangulated by a subcomplex $L \subset K$ and E is triangulated by a subcomplex $L_0 \subset L$. There exists a subdivision K' of K such that L_0 is full in K'. Let L' be the induced subdivision of L, $L' \subset K'$. There exists a subdivision K'' of K' such that L' is full in K''. As L_0 is full in K', it is also full in K''. Since $L' \subset K''$, L_0 is full in L'. We conclude that we may assume that the triple (X, X_k, E) is triangulated by (K, L, L_0) such that L_0 is full in L, L_0 is full in K, and L is full in K. Let K' be a first derived of K, inducing

L_0', L', so that the derived neighborhood $|N(L_0, L')|$ is a regular neighborhood of E in X_k, $|N(L, K')|$ is a regular neighborhood of X_k in X, and $|N(L_0, K')|$ is a regular neighborhood of E in X.

Step 2. Blocking $|N(L_0, L')|$ and fixing a projection to E : Note that

$$|N(L_0, L')| = \bigcup \{|\sigma^*(L')| : \sigma \in L_0\}.$$

Consider the set $B(\xi) = \{|\sigma^*(L_0')| : \sigma \in L_0\}$. Since E is a manifold, $B(\xi)$ is in fact a cell complex which covers E. To each $|\sigma^*(L_0')| \in B(\xi)$ associate the dual block $|\sigma^*(L')|$. It follows from Rourke and Sanderson's work [**RS68**] that the collection $\xi = \{|\sigma^*(L')| : \sigma \in L_0\}$ defines a block bundle $\xi/B(\xi)$ over the base complex $B(\xi)$ (this is where we are using that $E \hookrightarrow X_k$ is locally flat.) ξ has total space $E(\xi) = |N(L_0, L')|$. Since $E \hookrightarrow X_k$ is a good embedding (and using the above lemma 3.1), there exists a projection $p_1 : E(\xi) \longrightarrow E$ making ξ into a block fibration. The fiber of p_1 is of course a $(k-l)$-disc D^{k-l}.

Step 3. Blocking $|N(L, K')|$ and fixing a projection to X_k : We do this in parallel to the previous step. We have

$$|N(L, K')| = \bigcup \{|\sigma^*(K')| : \sigma \in L\}.$$

Consider $B(\eta) = \{|\sigma^*(L')| : \sigma \in L\}$. Since X_k is a manifold, $B(\eta)$ is a cell complex which covers X_k. To each $|\sigma^*(L')| \in B(\eta)$ associate the dual block $|\sigma^*(K')|$. As X^n is a PL-variety, it follows from D. Stone's work [**Sto72**] that the collection $\eta = \{|\sigma^*(K')| : \sigma \in L\}$ defines a cone block bundle $\eta/B(\eta)$ over the base complex $B(\eta)$. η has total space $E(\eta) = |N(L, K')|$. By the hypothesis on X and lemma 3.2, there exists a projection $p_2 : E(\eta) \longrightarrow X_k$ making η into a block fibration. The fiber of p_2 is the cone on some compact polyhedron F.

Step 4. Blocking $|N(L_0, K')|$ and establishing its structure as a block fibration with fiber $c(S^{k-l-1} * F)$, where $*$ denotes the join operation:
$|N(L_0, K')| = \bigcup \{|\sigma^*(K')| : \sigma \in L_0\}$, so consider $\zeta = \{|\sigma^*(K')| : \sigma \in L_0\}$ and $B(\zeta) = B(\xi)$. We will verify that ζ is a block bundle and $E(\zeta) \xrightarrow{p_1 p_2} E$ a block fibration. Given $|\sigma^*(L_0')| \in B(\zeta)$, we have to produce a structure ϕ_0 for the associated block $|\sigma^*(K')|$ which commutes with the projection $p_1 p_2$:

$$\begin{array}{ccc} |\sigma^*(K')| & \xrightarrow{\phi_0} & D^i \times c(S^{k-l-1} * F) \\ \Big\downarrow {\scriptstyle (p_1 p_2)|} & & \Big\downarrow {\scriptstyle \text{proj}} \\ |\sigma^*(L_0')| & \xrightarrow{\phi_0|} & D^i \end{array}$$

where $i = \dim \sigma^*(L'_0)$. Since ξ is a block fibration, there is a structure

$$\begin{array}{ccc} |\sigma^*(L')| & \xrightarrow{\psi} & D^i \times D^{k-l} \\ \Big\downarrow{\scriptstyle p_1|} & & \Big\downarrow{\scriptstyle \mathrm{proj}} \\ |\sigma^*(L'_0)| & \xrightarrow{\psi|} & D^i \end{array}$$

and since η is a block fibration, there is a structure

$$\begin{array}{ccc} |\sigma^*(K')| & \xrightarrow{\phi} & D^{i+k-l} \times cF \\ \Big\downarrow{\scriptstyle p_2|} & & \Big\downarrow{\scriptstyle \mathrm{proj}} \\ |\sigma^*(L')| & \xrightarrow{\phi|} & D^{i+k-l} \end{array}$$

Let ϕ_0 be the composition

$$|\sigma^*(K')| \xrightarrow{\phi} D^{i+k-l} \times cF \xrightarrow{(\phi|^{-1},1)} |\sigma^*(L')| \times cF$$

$$\xrightarrow{(\psi,1)} D^i \times D^{k-l} \times cF \xrightarrow{\cong} D^i \times c(S^{k-l-1} * F).$$

The following commutative diagram shows that ϕ_0 indeed works:
(write $\bar{\sigma} = |\sigma^*(L'_0)|$.)

$$\begin{array}{ccccc} |\sigma^*(K')| & \xrightarrow{p_2|} & |\sigma^*(L')| & \xrightarrow{p_1|} & \bar{\sigma} \\ \Big\downarrow{\scriptstyle \phi} & & \Big\downarrow{\scriptstyle \phi|} & & \Big\downarrow{\scriptstyle \phi|} \\ D^{i+k-l} \times cF & \xrightarrow{\mathrm{proj}} & D^{i+k-l} & & \phi(\bar{\sigma}) \\ \Big\downarrow{\scriptstyle (\phi|^{-1},1)} & & \Big\downarrow{\scriptstyle \phi|^{-1}} & & \Big\downarrow{\scriptstyle \phi|^{-1}} \\ |\sigma^*(L')| \times cF & \xrightarrow{\mathrm{proj}} & |\sigma^*(L')| & \xrightarrow{p_1|} & \bar{\sigma} \\ \Big\downarrow{\scriptstyle (\psi,1)} & & \Big\downarrow{\scriptstyle \psi} & & \Big\downarrow{\scriptstyle \psi|} \\ D^i \times D^{k-l} \times cF & \xrightarrow{\mathrm{proj}} & D^i \times D^{k-l} & \xrightarrow{\mathrm{proj}} & D^i \\ \Big\downarrow{\scriptstyle \cong} & & & & \Big\| \\ D^i \times c(S^{k-l-1} * F) & & \xrightarrow{\mathrm{proj}} & & D^i \end{array}$$

Summarizing, we have seen that a regular neighborhood of E in X is the total space $E(\zeta)$ of a block fibration $E(\zeta) \xrightarrow{p} E$, where we write $p = p_1 p_2$.

Step 5. Blocking $E(\zeta) \xrightarrow{\pi p} B$: We are headed towards showing that $E(\zeta) \xrightarrow{\pi p} B$ is a block fibration. In the present step 5 we decompose $E(\zeta)$ into polyhedra, the collection of which will be seen (in step 7) to form a block bundle over B. The structure isomorphisms for the blocks will commute with πp, establishing $E(\zeta) \xrightarrow{\pi p} B$ as a block fibration. The intermediate step 6 serves to describe the fiber of πp. $E \xrightarrow{\pi} B$ is a block bundle $\alpha/B(\alpha)$, where $B(\alpha)$ is a cell complex such that $|B(\alpha)| = B$. So E is covered by the blocks of α, but also by the cells in $B(\zeta)$. Triangulate E by a simplicial complex Q such that every block of α is triangulated by a subcomplex of Q. Recall that E is also triangulated by L_0' such that every cell in $B(\zeta)$ is covered by a subcomplex of L_0'. Let Q' be a common subdivision of Q and L_0'. Define the subdivision ζ'/Q' by $\zeta' = \{p^{-1}(|\sigma|) : \sigma \in Q'\}$. For $\Delta \in B(\alpha)$, write $\alpha(\Delta)$ for the block in α over Δ. Let Q'_Δ be the subcomplex of Q' which triangulates $|\alpha(\Delta)|$. Define

$$\gamma(\Delta) = \bigcup \{|\zeta'(\sigma)| : \sigma \in Q'_\Delta\},$$

where $\zeta'(\sigma)$ denotes the block in ζ' over σ, i.e. $|\zeta'(\sigma)| = p^{-1}(|\sigma|)$. In other words, $\gamma(\Delta) = p^{-1}(|Q'_\Delta|)$. We claim that the collection $\gamma = \{\gamma(\Delta) : \Delta \in B(\alpha)\}$ defines a block bundle $\gamma/B(\alpha)$. Before giving structures for the polyhedra in γ and thereby proving them to be blocks, we describe the fiber of πp.

Step 6. The fiber of πp has constant isomorphism type over B : Fix a basepoint $b_0 \in B$. Let $P = p^{-1}(\pi^{-1}(b_0))$ be the fiber over b_0. Let $A = \{x \in B : p^{-1}(\pi^{-1}(x)) \cong P\}$; we will show the following: If $x \in A$ and $\Delta \in B(\alpha)$ with $x \in |\Delta|$, then $|\Delta| \subset A$. It follows from the connectedness of B that the fiber of πp at every point of B is isomorphic to P.
So let $x \in A$, $\Delta \in B(\alpha)$, $x \in |\Delta|$. The block $\alpha(\Delta)$ has a structure ϕ,

$$\begin{array}{ccc} |\alpha(\Delta)| & \xrightarrow{\phi} & D^m \times F_0 \\ \Big\downarrow {\scriptstyle \pi|} & & \Big\downarrow {\scriptstyle \mathrm{proj}} \\ |\Delta| & \xrightarrow{\phi|} & D^m \end{array}$$

where $m = \dim \Delta$. Let $\zeta'|$ be the restriction of ζ' to $|\alpha(\Delta)|$. Let $\zeta_0 = (\phi^{-1})^*(\zeta'|)$ be the block fibration $E(\zeta_0) \xrightarrow{p_0} D^m \times F_0$ induced by ϕ^{-1}. As ζ_0 is a bundle over a polyhedron cross a disc, it trivializes along the disc factor, that is, there is an isomorphism

$$\zeta_0 \cong D^m \times (\zeta_0|_{\{\phi(x)\} \times F_0})$$

inducing

$$\begin{array}{ccc} E(\zeta_0) & \xrightarrow{\cong} & D^m \times E(\zeta_0|_{\{\phi(x)\}\times F_0}) \\ \Big\downarrow p_0 & & \Big\downarrow (1,p_0|) \\ D^m \times F_0 & = & D^m \times F_0, \end{array}$$

compare theorem 1.11 [**RS68**] or theorem I, 1.11 [**Sto72**]. From the commutative diagram

$$\begin{array}{ccccc} E(\zeta'|) & \xrightarrow{\cong} & E(\zeta_0) & \xrightarrow{\cong} & D^m \times E(\zeta_0|_{\{\phi(x)\}\times F_0}) \\ \Big\downarrow p| & & \Big\downarrow p_0 & & \Big\downarrow (1,p_0|) \\ |\alpha(\Delta)| & \xrightarrow{\phi} & D^m \times F_0 & = & D^m \times F_0 \\ \Big\downarrow \pi| & & \Big\downarrow \mathrm{proj} & & \Big\downarrow \mathrm{proj} \\ |\Delta| & \xrightarrow{\phi|} & D^m & = & D^m \end{array}$$

we see that if $y \in |\Delta|$, then

$$p^{-1}(\pi^{-1}(y)) \cong (1,p_0|)^{-1}\mathrm{proj}^{-1}(\phi(y)) = (1,p_0|)^{-1}\{\phi(y)\} \times F_0 =$$

$$\{\phi(y)\} \times E(\zeta_0|_{\{\phi(x)\}\times F_0}) \cong E(\zeta'|\pi^{-1}(x)) = p^{-1}(\pi^{-1}(x)) \cong P.$$

Therefore, $y \in A$ and $|\Delta| \subset A$.

Step 7. $E(\zeta) \xrightarrow{\pi p} B$ is a block fibration with respect to γ : From the preceding step it is clear how to obtain a structure for the block $\gamma(\Delta) \in \gamma$. Identifying $|\gamma(\Delta)| = E(\zeta'|)$, the composition

$$\psi : |\gamma(\Delta)| = E(\zeta'|) \xrightarrow{\cong} E(\zeta_0) \xrightarrow{\cong} D^m \times E(\zeta_0|_{\{\phi(x)\}\times F_0}) \cong D^m \times P$$

is a structure

$$\begin{array}{ccc} |\gamma(\Delta)| & \xrightarrow{\psi} & D^m \times P \\ \Big\downarrow (\pi p)| & & \Big\downarrow \mathrm{proj} \\ |\Delta| & \xrightarrow{\psi|} & D^m \end{array}$$

Step 8. The boundary of $E(\zeta)$: Since ζ' is a cone block bundle, the rim of a block $\zeta'(\sigma), \sigma \in Q'$, is well defined, see [**Sto72**]. In fact, let $\mathrm{rim}\zeta'(\sigma) = \phi^{-1}(D^m \times (S^{k-l-1} * F))$, where ϕ is a structure for $\zeta'(\sigma)$ and $m = \dim\sigma$ (recall that the fiber of ζ' is $c(S^{k-l-1} * F)$.) Define ζ^r to be the collection of all rims of blocks in ζ',

$$\zeta^r = \{\mathrm{rim}\zeta'(\sigma) : \sigma \in Q'\},$$

and let $E(\zeta^r) = \bigcup\{\text{rim} \in \zeta^r\}$. Then $\partial E(\zeta) = E(\zeta^r)$. Let p^r be the restriction of p to $E(\zeta^r)$. Then $E(\zeta^r) \xrightarrow{p^r} E$ is a fibration with fiber $S^{k-l-1} * F$, however it is technically not a block bundle since it does not have a preferred zero section. Composing with π, we see that $E(\zeta^r) \xrightarrow{\pi p^r} B$ is a fibration with fiber ∂P.

Step 9. The mapping cylinder of πp^r : Triangulate $E(\zeta^r)$ by K and B by L such that $\pi p^r : |K| \longrightarrow |L|$ is simplicial. Since $E(\zeta^r)$ is a compact polyhedron, K is a finite complex, so πp^r is a proper simplicial map. For any proper simplicial map, the simplicial mapping cylinder is defined, see [**Ver84**] or [**Coh67**]. Denote the simplicial mapping cylinder of πp^r by M_s. M_s comes with a canonical retraction $r_s : |M_s| \longrightarrow B$ (which is simplicial for suitable subdivisions of M_s and L.)

REMARK 3.1. If M_{top} is the topological mapping cylinder of πp^r and $r : M_{top} \to B$ the canonical retraction, then there exists a homeomorphism $|M_s| \xrightarrow{\cong} M_{top}$ such that

$$\begin{array}{ccc} |M_s| & \xrightarrow{\cong} & M_{top} \\ \downarrow{\scriptstyle r_s} & & \downarrow{\scriptstyle r} \\ B & = & B \end{array}$$

commutes.

Subdivide M_s into M_s' and L into L' such that $r_s : |M_s'| \to |L'|$ is a simplicial map. Triangulate B by a simplicial complex Q such that every cell in $B(\alpha)$ is covered by a subcomplex of Q. Let Q' be a common subdivision of Q and L'. There exists a subdivision M_s'' of M_s' such that $r_s : |M_s''| \to |Q'|$ is simplicial. Consider $\mu = \{r_s^{-1}(|\Delta|) : \Delta \in Q'\}$. We claim that μ/Q' is a block bundle. Let γ'/Q' be the subdivision $\gamma' = \{(\pi p)^{-1}(|\Delta|) : \Delta \in Q'\}$ of $\gamma/B(\alpha)$. Let $\Delta \in Q'$. The block $(\pi p)^{-1}(|\Delta|) \in \gamma'$ has a structure

$$\begin{array}{ccc} (\pi p)^{-1}(|\Delta|) & \xrightarrow{\phi} & D^m \times P \\ \downarrow{\scriptstyle (\pi p)|} & & \downarrow{\scriptstyle \text{proj}} \\ |\Delta| & \xrightarrow{\phi|} & D^m \end{array}$$

Restricting ϕ to the rim of the blocks over $\pi^{-1}(|\Delta|)$, we get

$$\begin{array}{ccc} (\pi p^r)^{-1}(|\Delta|) & \xrightarrow{\phi|} & D^m \times \partial P \\ \downarrow{\scriptstyle (\pi p^r)|} & & \downarrow{\scriptstyle \text{proj}} \\ |\Delta| & \xrightarrow{\phi|} & D^m \end{array}$$

Write $f = (\pi p^r)|$ for the restriction of πp^r to $(\pi p^r)^{-1}(|\Delta|)$. As the mapping cylinder is a functorial construction, $\phi|$ induces a simplicial map $\Phi : |M_s(f)| \xrightarrow{\cong} |M_s(\text{proj})|$ which extends $\phi|$ and such that

$$\begin{array}{ccc} |M_s(f)| & \xrightarrow{\Phi} & |M_s(\text{proj})| \\ \Big\downarrow{\scriptstyle r_s(f)} & & \Big\downarrow{\scriptstyle r_s(\text{proj})} \\ |\Delta| & \xrightarrow{\phi|} & D^m \end{array}$$

commutes, where $r_s(f)$ and $r_s(\text{proj})$ are the canonical retractions, see lemma 9.4.3 [**Ver84**].

Observe now that $|M_s(f)| = r_s^{-1}(|\Delta|)$ and $|M_s(\text{proj})| = D^m \times c(\partial P)$, so we have

$$\begin{array}{ccc} r_s^{-1}(|\Delta|) & \xrightarrow{\Phi} & D^m \times c(\partial P) \\ \Big\downarrow{\scriptstyle r_s|} & & \Big\downarrow{\scriptstyle \text{proj}} \\ |\Delta| & \xrightarrow{\Phi|=\phi|} & D^m \end{array}$$

So Φ is the desired structure for $r_s^{-1}(|\Delta|) \in \mu$. Therefore, $|M_s| \xrightarrow{r_s} B$ is in fact a block fibration with respect to μ.

Step 10. The construction of the PL-variety Y : We define Y by deleting from X a regular neighborhood of E in X and gluing in the mapping cylinder $|M_s|$ along the common boundary $E(\zeta^r)$.

$$Y = (X^n - \text{int}E(\zeta)) \cup_{E(\zeta^r)} |M_s|.$$

It follows from the previous step that Y is a PL-variety with respect to a filtration of the form

$$Y^n = Y_n \supset Y_{n-1} \supset Y_{n-2} \supset \ldots \supset Y_k \supset Y_b,$$

with $Y_b = B^b$, i.e. we have generated a new bottom stratum of dimension b. The link of the stratum Y_b in Y is ∂P. Also note that Y satisfies the assumption on X that we made at the outset of the construction: X was required to be such that a blocking of a regular neighborhood of the bottom stratum is a block fibration. Since $|M_s|$ was shown to be a block fibration with respect to r_s, Y again satisfies the requirement. This enables us to use the construction inductively.

4. Inductive Singularization of a Manifold

Start with a closed PL-manifold M^n, choose a block fibration $E_0 \xrightarrow{\pi_0} B^{b(0)}$, $E_0, B^{b(0)}$ closed manifolds, and a good embedding $f_0 : E_0 \hookrightarrow M$. Set

$$X^{(1)} = Y(M, \pi_0, f_0),$$

where $Y(M, \pi_0, f_0)$ denotes the space Y obtained from the given construction with input data M, π_0 and f_0. Suppose

$$X^{(m)} = X_n^{(m)} \supset X_{b(0)}^{(m)} \supset \ldots \supset X_{b(m-1)}^{(m)}$$

has been constructed. Choose a block fibration $E_m \xrightarrow{\pi_m} B^{b(m)}$ and a good embedding $f_m : E_m \hookrightarrow X^{(m)}_{b(m-1)}$. Set

$$X^{(m+1)} = Y(X^{(m)}, \pi_m, f_m).$$

After say m_0 steps, one arrives at a resulting space $Z = X^{(m_0)}$.

DEFINITION 3.1. The sequence $\{Z, X^{(m_0-1)}, X^{(m_0-2)}, \dots, X^{(1)}, M\}$ is called an *ordered resolution* of the space Z.

This kind of resolution facilitates the construction of a self-dual complex of sheaves on Z, as will be seen in chapter 5.

CHAPTER 4

The Cobordism Group Ω_*^{SD}

We set up a cobordism group Ω_*^{SD} whose elements are equivalence classes of closed pseudomanifolds supporting a self-dual sheaf. In particular, an appropriate notion of cobordism and boundary operator will be defined. A pair (pseudomanifold, self-dual sheaf) has a tautological signature associated to it, namely the signature of the quadratic form on hypercohomology in the middle dimension, induced by the self-duality isomorphism. We prove that this signature is a cobordism invariant. This result is crucial in proving our main theorem on ordered resolutions, see section 1 of chapter 5.
Moreover, there is a homomorphism from Witt space cobordism [**Sie83**] to Ω_*^{SD} :

$$\Omega_*^{\text{Witt}} \longrightarrow \Omega_*^{SD}$$

and our signature extends $\sigma : \Omega_*^{\text{Witt}} \to \mathbb{Z}$.
In the formal development of Ω_*^{SD} we follow to some extent Stong [**Sto68**].

1. The Closed Objects

Define $\mathcal{C}^n$ (the closed objects) to be the set of triples

$$\mathcal{C}^n = \{(X^n, \mathbf{A}^\bullet, d)\},$$

where X^n is an n-dimensional closed oriented pseudomanifold and $(\mathbf{A}^\bullet, d) \in SD(X)$. Disjoint union defines an operation

$$\mathcal{C}^n \times \mathcal{C}^n \xrightarrow{+} \mathcal{C}^n.$$

More precisely, if $(X_1, \mathbf{A}_1^\bullet, d_1), (X_2, \mathbf{A}_2^\bullet, d_2) \in \mathcal{C}^n$, then

$$(X_1, \mathbf{A}_1^\bullet, d_1) + (X_2, \mathbf{A}_2^\bullet, d_2) := (X_1 \cup X_2, \mathbf{S}^\bullet, d),$$

where $\mathbf{S}^\bullet|_{X_1} = \mathbf{A}_1^\bullet$, $\mathbf{S}^\bullet|_{X_2} = \mathbf{A}_2^\bullet$, $d|_{X_1} = d_1$, and $d|_{X_2} = d_2$.
We call $(X_1, \mathbf{A}_1^\bullet, d_1)$ and $(X_2, \mathbf{A}_2^\bullet, d_2)$ *isomorphic*,

$$(X_1, \mathbf{A}_1^\bullet, d_1) \cong (X_2, \mathbf{A}_2^\bullet, d_2),$$

if there exist a stratified homeomorphism $f : X_1 \to X_2$ and an isomorphism $g \in \operatorname{Hom}_{D^b(X_1)}(\mathbf{A}_1^\bullet, f^*\mathbf{A}_2^\bullet)$ such that

$$\begin{array}{ccc}
\mathcal{D}\mathbf{A}_1^\bullet[n] & \xleftarrow[\cong]{\mathcal{D}g[n]} & \mathcal{D}f^*\mathbf{A}_2^\bullet[n] \cong f^*\mathcal{D}\mathbf{A}_2^\bullet[n] \\
{\scriptstyle d_1}\downarrow\cong & & \cong\downarrow{\scriptstyle f^*d_2} \\
\mathbf{A}_1^\bullet & \xrightarrow[\cong]{g} & f^*\mathbf{A}_2^\bullet
\end{array}$$

commutes.

Given $(X^{2k}, \mathbf{A}^\bullet, d)$, d induces a nonsingular pairing

$$\mathcal{H}^{-k}(X; \mathbf{A}^\bullet) \otimes \mathcal{H}^{-k}(X; \mathbf{A}^\bullet) \longrightarrow \mathbb{R}.$$

Let $\sigma(X, \mathbf{A}^\bullet, d)$ denote the signature of this pairing and set $\sigma(X^n, \mathbf{A}^\bullet, d) = 0$ for n odd. This defines a map

$$\sigma : \mathcal{C}^n \longrightarrow \mathbb{Z}.$$

Clearly, σ has the properties

$$\begin{aligned} &\sigma(X_1, \mathbf{A}_1^\bullet, d_1) = \sigma(X_2, \mathbf{A}_2^\bullet, d_2) \text{ if } (X_1, \mathbf{A}_1^\bullet, d_1) \cong (X_2, \mathbf{A}_2^\bullet, d_2), \\ &\sigma((X_1, \mathbf{A}_1^\bullet, d_1) + (X_2, \mathbf{A}_2^\bullet, d_2)) = \sigma(X_1, \mathbf{A}_1^\bullet, d_1) + \sigma(X_2, \mathbf{A}_2^\bullet, d_2). \end{aligned}$$

2. The Admissible Cobordisms

Define Cob^{n+1} (the admissible cobordisms) to be the set of triples

$$Cob^{n+1} = \{(Y^{n+1}, \mathbf{B}^\bullet, \delta)\},$$

where Y^{n+1} is an $(n+1)$-dimensional compact oriented pseudomanifold with boundary, $(\mathbf{B}^\bullet, \delta) \in SD(\operatorname{int} Y)$, $\delta : \mathcal{D}\mathbf{B}^\bullet[n+1] \xrightarrow{\cong} \mathbf{B}^\bullet$. Again, disjoint sum defines an operation

$$Cob^{n+1} \times Cob^{n+1} \xrightarrow{+} Cob^{n+1}.$$

Let i, j denote the inclusions $i : \operatorname{int} Y \hookrightarrow Y$, $j : \partial Y \hookrightarrow Y$.

LEMMA 4.1. *If $\boldsymbol{B}^\bullet \in D^b(\operatorname{int} Y)$, then there is a natural isomorphism*

$$j^* Ri_* \boldsymbol{B}^\bullet \xrightarrow{\cong} j^! Ri_! \boldsymbol{B}^\bullet[1]$$

PROOF. Consider the distinguished triangle ($\mathbf{A}^\bullet \in D^b(Y)$)

$$\begin{array}{ccc} Ri_! i^* \mathbf{A}^\bullet & \longrightarrow & \mathbf{A}^\bullet \\ & {\scriptstyle [1]}\nwarrow \quad \swarrow & \\ & j_* j^* \mathbf{A}^\bullet & \end{array}$$

Apply j^* :

$$\begin{array}{ccc} j^* Ri_! i^* \mathbf{A}^\bullet & \longrightarrow & j^* \mathbf{A}^\bullet \\ & {\scriptstyle [1]}\nwarrow \quad \swarrow{\scriptstyle \cong} & \\ & j^* \mathbf{A}^\bullet & \end{array}$$

It follows that $j^* Ri_! i^* \mathbf{A}^\bullet \cong 0$. Substituting $\mathbf{A}^\bullet = Ri_* \mathbf{B}^\bullet$ and using $i^* Ri_* \mathbf{B}^\bullet \cong \mathbf{B}^\bullet$, we see

$$j^* Ri_! \mathbf{B}^\bullet \cong 0.$$

In the triangle

$$\begin{array}{ccc} j^!\mathbf{A}^\bullet & \longrightarrow & j^*\mathbf{A}^\bullet \\ & {\scriptstyle [1]}\nwarrow \qquad \swarrow & \\ & j^*Ri_*i^*\mathbf{A}^\bullet & \end{array}$$

set $\mathbf{A}^\bullet = Ri_!\mathbf{B}^\bullet$. We obtain

$$\begin{array}{ccc} j^!Ri_!\mathbf{B}^\bullet & \longrightarrow & j^*Ri_!\mathbf{B}^\bullet \cong 0 \\ & {\scriptstyle [1]}\nwarrow \qquad \swarrow & \\ & j^*Ri_*i^*Ri_!\mathbf{B}^\bullet \cong j^*Ri_*\mathbf{B}^\bullet & \end{array}$$

and hence $j^*Ri_*\mathbf{B}^\bullet \xrightarrow{\cong} j^!Ri_!\mathbf{B}^\bullet[1]$. □

Suppose we are given $(Y^{n+1}, \mathbf{B}^\bullet, \delta) \in Cob^{n+1}$, $\delta : \mathcal{D}\mathbf{B}^\bullet[n+1] \xrightarrow{\cong} \mathbf{B}^\bullet$. An application of the functor j^*Ri_* produces an induced isomorphism

$$j^*Ri_*(\delta) : j^*Ri_*\mathcal{D}\mathbf{B}^\bullet[n+1] \xrightarrow{\cong} j^*Ri_*\mathbf{B}^\bullet.$$

We have

$$j^*Ri_*\mathcal{D}\mathbf{B}^\bullet[n+1] \cong \mathcal{D}(j^!Ri_!\mathbf{B}^\bullet)[n+1]$$

and according to lemma 4.1,

$$j^*Ri_*\mathbf{B}^\bullet \cong j^!Ri_!\mathbf{B}^\bullet[1].$$

Hence δ induces a self-duality isomorphism d for $j^!Ri_!\mathbf{B}^\bullet$:

$$d : \mathcal{D}(j^!Ri_!\mathbf{B}^\bullet)[n] \xrightarrow{\cong} j^!Ri_!\mathbf{B}^\bullet$$

We call d the *boundary of* δ and write $d = \partial\delta$. In this fashion one defines a boundary map

$$\partial : Cob^{n+1} \longrightarrow \mathcal{C}^n$$

$$(Y^{n+1}, \mathbf{B}^\bullet, \delta) \overset{\partial}{\mapsto} (\partial Y, j^!Ri_!\mathbf{B}^\bullet, \partial\delta)$$

The property $\partial((Y_1, \mathbf{B}_1^\bullet, \delta_1) + (Y_2, \mathbf{B}_2^\bullet, \delta_2)) = \partial(Y_1, \mathbf{B}_1^\bullet, \delta_1) + \partial(Y_2, \mathbf{B}_2^\bullet, \delta_2)$ is obvious.

DEFINITION 4.1. Two triples $(X_1, \mathbf{A}_1^\bullet, d_1), (X_2, \mathbf{A}_2^\bullet, d_2) \in \mathcal{C}^n$ are *cobordant* if there exist $(Y_1, \mathbf{B}_1^\bullet, \delta_1), (Y_2, \mathbf{B}_2^\bullet, \delta_2) \in Cob^{n+1}$ such that

$$(X_1, \mathbf{A}_1^\bullet, d_1) + \partial(Y_1, \mathbf{B}_1^\bullet, \delta_1) \cong (X_2, \mathbf{A}_2^\bullet, d_2) + \partial(Y_2, \mathbf{B}_2^\bullet, \delta_2).$$

In this case we write $(X_1, \mathbf{A}_1^\bullet, d_1) \sim (X_2, \mathbf{A}_2^\bullet, d_2)$.

It is easily verified that $\sim$ is an equivalence relation. Write $[(X, \mathbf{A}^\bullet, d)]$ for the equivalence class of $(X, \mathbf{A}^\bullet, d) \in \mathcal{C}^n$.

Define

$$\Omega_n^{SD} := \{[(X, \mathbf{A}^\bullet, d)] : (X, \mathbf{A}^\bullet, d) \in \mathcal{C}^n\}$$

and

$$[(X_1, \mathbf{A}_1^\bullet, d_1)] + [(X_2, \mathbf{A}_2^\bullet, d_2)] := [(X_1, \mathbf{A}_1^\bullet, d_1) + (X_2, \mathbf{A}_2^\bullet, d_2)].$$

(we shall often simply write X for a triple $(X, \mathbf{A}^\bullet, d) \in \mathcal{C}^n$.) This operation is well defined: Suppose $X_1 \sim X_1'$ and $X_2 \sim X_2'$. Thus there exist $Y_1, Y_1', Y_2, Y_2' \in Cob^{n+1}$ such that $X_1 + \partial Y_1 \cong X_1' + \partial Y_1'$ and $X_2 + \partial Y_2 \cong X_2' + \partial Y_2'$. Then $(X_1 + X_2) + \partial(Y_1 + Y_2) \cong (X_1' + X_2') + \partial(Y_1' + Y_2')$ and $X_1 + X_2 \sim X_1' + X_2'$.

LEMMA 4.2. $[(X, \boldsymbol{A}^\bullet, d)] + [(X, \boldsymbol{A}^\bullet, -d)] = 0.$

PROOF. Let $Y = X \times [0,1]$. Then $\operatorname{int} Y = X \times (0,1)$ and we have the projection $\pi : X \times (0,1) \to X$. Set $\mathbf{B}^\bullet = \pi^! \mathbf{A}^\bullet$, $\mathbf{B}^\bullet \in D^b(\operatorname{int} Y)$. We show that $\mathbf{B}^\bullet$ is self-dual: By [**KS90**, prop. 3.3.2] we have an isomorphism

$$\pi^! \mathbf{A}^\bullet \cong \pi^! \mathbb{R}_X \otimes \pi^* \mathbf{A}^\bullet,$$

while for the constant sheaf

$$\pi^! \mathbb{R}_X \cong \pi^* \mathbb{R}_X[1],$$

hence

$$\pi^! \mathbf{A}^\bullet \cong \pi^* \mathbf{A}^\bullet[1].$$

Therefore, d induces

$$\delta : \mathcal{D}\mathbf{B}^\bullet[n+1] = \mathcal{D}(\pi^! \mathbf{A}^\bullet)[n+1] \cong \pi^* \mathcal{D}\mathbf{A}^\bullet[n+1] \overset{\pi^* d[1]}{\longrightarrow} \pi^* \mathbf{A}^\bullet[1] = \pi^! \mathbf{A}^\bullet = \mathbf{B}^\bullet$$

We have constructed an element $(Y, \mathbf{B}^\bullet, \delta) \in Cob^{n+1}$ with

$$\partial(Y, \mathbf{B}^\bullet, \delta) \cong (X, \mathbf{A}^\bullet, d) + (X, \mathbf{A}^\bullet, -d),$$

and so

$$(X, \mathbf{A}^\bullet, d) + (X, \mathbf{A}^\bullet, -d) + \partial\varnothing \cong \varnothing + \partial Y.$$

□

The lemma shows that the operation $+$ makes Ω_n^{SD} into an abelian group.

3. The Cobordism Invariance of σ

The following theorem implies the cobordism invariance of σ. It will be applied in section 6, chapter 5, to show that a certain subsheaf is Lagrangian.

THEOREM 4.1. *Given $(Y^{n+1}, \boldsymbol{B}^\bullet, \delta) \in Cob^{n+1}$, where $n = 2s$, let $\boldsymbol{E}^\bullet = Ri_! \boldsymbol{B}^\bullet$ and consider the canonical morphism $Ri_* i^* \boldsymbol{E}^\bullet \to j_* j^! \boldsymbol{E}^\bullet[1]$. The image*

$$\operatorname{Im}(\mathcal{H}^{-s-1}(Y; Ri_* i^* \boldsymbol{E}^\bullet) \to \mathcal{H}^{-s}(Y; j_* j^! \boldsymbol{E}^\bullet))$$

is a Lagrangian subspace in $\mathcal{H}^s(Y; j_* j^! \boldsymbol{E}^\bullet) \cong \mathcal{H}^s(\partial Y; j^! \boldsymbol{E}^\bullet)$.

PROOF. We have

$$\delta : \mathcal{D}\mathbf{B}^\bullet[n+1] \xrightarrow{\cong} \mathbf{B}^\bullet$$

and

$$\partial\delta : \mathcal{D}(j^! \mathbf{E}^\bullet)[n] \xrightarrow{\cong} j^! \mathbf{E}^\bullet$$

Since $j_*j^! \to 1$ is a natural transformation of functors, we have a commutative diagram

$$\begin{array}{ccc} j_*j^!(Ri_!\mathbf{B}^\bullet) & \longrightarrow & Ri_!\mathbf{B}^\bullet \\ {\scriptstyle j_*j^!Ri_!\delta}\uparrow & & \uparrow{\scriptstyle Ri_!\delta} \\ j_*j^!(Ri_!\mathcal{D}\mathbf{B}^\bullet[n+1]) & \longrightarrow & Ri_!\mathcal{D}\mathbf{B}^\bullet[n+1] \end{array}$$

Now

$$Ri_!\mathcal{D}\mathbf{B}^\bullet[n+1] \cong Ri_!\mathcal{D}(i^*\mathbf{E}^\bullet)[n+1] \cong Ri_!i^*\mathcal{D}\mathbf{E}^\bullet[n+1]$$

and using lemma 4.1,

$$j^!Ri_!\mathcal{D}\mathbf{B}^\bullet[n+1] \cong \mathcal{D}(j^*Ri_*\mathbf{B}^\bullet)[n+1] \cong \mathcal{D}(j^!Ri_!\mathbf{B}^\bullet[1])[n+1] \cong \mathcal{D}(j^!Ri_!\mathbf{B}^\bullet)[n]$$

Therefore, we get a commutative diagram

$$\begin{array}{ccc} j_*j^!\mathbf{E}^\bullet & \longrightarrow & \mathbf{E}^\bullet \\ {\scriptstyle j_*\partial\delta}\uparrow & & \uparrow{\scriptstyle Ri_!\delta} \\ j_*j^*\mathcal{D}\mathbf{E}^\bullet[n] & \longrightarrow & Ri_!i^*\mathcal{D}\mathbf{E}^\bullet[n+1] \end{array} \tag{27}$$

We show that the given data give rise to an isomorphism between the triangle

$$\begin{array}{ccc} j_*j^!\mathbf{E}^\bullet & \longrightarrow & \mathbf{E}^\bullet \\ {\scriptstyle [1]}\nwarrow & & \swarrow \\ & Ri_*i^*\mathbf{E}^\bullet & \end{array}$$

and the triangle

$$\begin{array}{ccc} Ri_!i^*\mathcal{D}\mathbf{E}^\bullet[n+1] & \longrightarrow & \mathcal{D}\mathbf{E}^\bullet[n+1] \\ {\scriptstyle [1]}\nwarrow & & \swarrow \\ & j_*j^*\mathcal{D}\mathbf{E}^\bullet[n+1] & \end{array}$$

Consider the composition

$$\mathcal{D}\mathbf{E}^\bullet[n+1] \cong \mathcal{D}(Ri_!\mathbf{B}^\bullet)[n+1] \cong Ri_*\mathcal{D}\mathbf{B}^\bullet[n+1] \cong Ri_*\mathbf{B}^\bullet \cong Ri_*i^*\mathbf{E}^\bullet$$

and the composition

$$Ri_!i^*\mathcal{D}\mathbf{E}^\bullet[n+1] \cong Ri_!\mathcal{D}(i^*\mathbf{E}^\bullet)[n+1] \cong Ri_!\mathcal{D}\mathbf{B}^\bullet[n+1] \cong Ri_!\mathbf{B}^\bullet \cong \mathbf{E}^\bullet.$$

These fit into a commutative square

$$\begin{array}{ccc}
\mathbf{E}^\bullet & \longrightarrow & Ri_*i^*\mathbf{E}^\bullet \\
\uparrow \cong & & \uparrow \cong \\
Ri_!i^*\mathcal{D}\mathbf{E}^\bullet[n+1] & \longrightarrow & \mathcal{D}\mathbf{E}^\bullet[n+1]
\end{array}$$

Applying the dualizing functor to diagram (27) one obtains

$$\begin{array}{ccc}
\mathcal{D}\mathbf{E}^\bullet & \longrightarrow & j_*j^*\mathcal{D}\mathbf{E}^\bullet \\
\downarrow \cong & & \downarrow \cong \\
Ri_*i^*\mathbf{E}^\bullet[-n-1] & \longrightarrow & j_*j^!\mathbf{E}^\bullet[-n]
\end{array}$$

Shifting the square by $n+1$ yields

$$\begin{array}{ccc}
\mathcal{D}\mathbf{E}^\bullet[n+1] & \longrightarrow & j_*j^*\mathcal{D}\mathbf{E}^\bullet[n+1] \\
\downarrow \cong & & \downarrow \cong \\
Ri_*i^*\mathbf{E}^\bullet & \longrightarrow & j_*j^!\mathbf{E}^\bullet[1]
\end{array}$$

Putting our squares together, we obtain the isomorphism between the two triangles:

$$\begin{array}{ccccccc}
j_*j^!\mathbf{E}^\bullet & \longrightarrow & \mathbf{E}^\bullet & \longrightarrow & Ri_*i^*\mathbf{E}^\bullet & \longrightarrow & j_*j^!\mathbf{E}^\bullet[1] \\
\uparrow \cong & & \uparrow \cong & & \uparrow \cong & & \uparrow \cong \\
j_*j^*\mathcal{D}\mathbf{E}^\bullet[n] & \longrightarrow & Ri_!i^*\mathcal{D}\mathbf{E}^\bullet[n+1] & \longrightarrow & \mathcal{D}\mathbf{E}^\bullet[n+1] & \longrightarrow & j_*j^*\mathcal{D}\mathbf{E}^\bullet[n+1]
\end{array}$$

Taking hypercohomology, we compute using Borel-Moore duality

$$\mathcal{H}^k(j_*j^!\mathbf{E}^\bullet) \cong \mathcal{H}^k(j_*j^*\mathcal{D}\mathbf{E}^\bullet[n])$$

$$\cong \mathcal{H}^{k+n}(\mathcal{D}(j_*j^!\mathbf{E}^\bullet)) \cong \mathrm{Hom}(\mathcal{H}^{-k-n}(j_*j^!\mathbf{E}^\bullet), \mathbb{R}),$$

$$\mathcal{H}^k(\mathbf{E}^\bullet) \cong \mathcal{H}^k(Ri_!i^*\mathcal{D}\mathbf{E}^\bullet[n+1]) \cong \mathcal{H}^{k+n+1}(Ri_!i^*\mathcal{D}\mathbf{E}^\bullet)$$

$$\cong \mathcal{H}^{k+n+1}(\mathcal{D}(Ri_*i^*\mathbf{E}^\bullet)) \cong \mathrm{Hom}(\mathcal{H}^{-k-n-1}(Ri_*i^*\mathbf{E}^\bullet), \mathbb{R}),$$

and

$$\mathcal{H}^k(Ri_*i^*\mathbf{E}^\bullet) \cong \mathcal{H}^k(\mathcal{D}\mathbf{E}^\bullet[n+1]) \cong \mathcal{H}^{k+n+1}(\mathcal{D}\mathbf{E}^\bullet)$$

$$\cong \mathrm{Hom}(\mathcal{H}^{-k-n-1}(\mathbf{E}^\bullet), \mathbb{R}).$$

For $k = -s$ we have $-k - n = -s$. Therefore, the long exact hypercohomology sequence associated with the first triangle is dually paired to itself:

$$\begin{array}{ccccccc}
\to \mathcal{H}^{-s-1}(\mathbf{E}^\bullet) & \to & \mathcal{H}^{-s-1}(Ri_*i^*\mathbf{E}^\bullet) & \to & \mathcal{H}^{-s}(j_*j^!\mathbf{E}^\bullet) & \to & \mathcal{H}^{-s}(\mathbf{E}^\bullet) \\
\otimes & & \otimes & & \otimes & & \otimes \\
\leftarrow \mathcal{H}^{-s}(Ri_*i^*\mathbf{E}^\bullet) & \leftarrow & \mathcal{H}^{-s}(\mathbf{E}^\bullet) & \leftarrow & \mathcal{H}^{-s}(j_*j^*\mathbf{E}^\bullet) & \leftarrow & \mathcal{H}^{-s-1}(Ri_*i^*\mathbf{E}^\bullet) \\
\downarrow & & \downarrow & & \downarrow & & \downarrow \\
\mathbb{R} & & \mathbb{R} & & \mathbb{R} & & \mathbb{R}
\end{array}$$

To produce the Lagrangian subspace L in $\mathcal{H}^{-s}(\partial Y; j^!\mathbf{E}^\bullet) \cong \mathcal{H}^{-s}(Y; j_*j^!\mathbf{E}^\bullet)$, take

$$L := \operatorname{image}(\mathcal{H}^{-s-1}(Ri_*i^*\mathbf{E}^\bullet) \to \mathcal{H}^s(j_*j^!\mathbf{E}^\bullet)).$$

□

COROLLARY 4.1. *If* $(Y, \boldsymbol{B}^\bullet, \delta) \in Cob^{n+1}$, *then*

$$\sigma(\partial(Y, \boldsymbol{B}^\bullet, \delta)) = 0.$$

COROLLARY 4.2. *If* $(X_i, \boldsymbol{A}_i^\bullet, d_i) \in \mathcal{C}^n$, $i = 1, 2$, *and* $(X_1, \boldsymbol{A}_1^\bullet, d_1) \sim (X_2, \boldsymbol{A}_2^\bullet, d_2)$, *then*

$$\sigma(X_1, \boldsymbol{A}_1^\bullet, d_1) = \sigma(X_2, \boldsymbol{A}_2^\bullet, d_2)$$

PROOF. There exist $Y_1, Y_2 \in Cob^{n+1}$ with $X_1 + \partial Y_1 \cong X_2 + \partial Y_2$. By corollary 4.1, $\sigma(\partial Y_1) = 0$, $\sigma(\partial Y_2) = 0$. We calculate:

$$\begin{aligned}
\sigma(X_1) &= \sigma(X_1) + \sigma(\partial Y_1) = \sigma(X_1 + \partial Y_1) \\
&= \sigma(X_2 + \partial Y_2) = \sigma(X_2) + \sigma(\partial Y_2) = \sigma(X_2).
\end{aligned}$$

□

Summarizing the above results, σ is a well defined group homomorphism

$$\sigma : \Omega_*^{SD} \longrightarrow \mathbb{Z},$$

where $\Omega_*^{SD} = \bigoplus_{n \geq 0} \Omega_n^{SD}$.

4. Relation to Witt Space Cobordism

We now turn to constructing a homomorphism

$$\Omega_*^{\text{Witt}} \longrightarrow \Omega_*^{SD}$$

Here Ω_*^{Witt} denotes the Witt cobordism group as defined (and computed) in [**Sie83**]. Let X^n be a closed oriented Witt space. The intersection chain complex $\mathbf{IC}_{\bar{m}}^\bullet(X)$ on X is self-dual, and the isomorphism

$$d_W : \mathcal{D}\mathbf{IC}_{\bar{m}}^\bullet(X)[n] \xrightarrow{\cong} \mathbf{IC}_{\bar{m}}^\bullet(X)$$

is uniquely determined by the orientation of X. We claim that the assignment

$$X^n \mapsto (X^n, \mathbf{IC}_{\bar{m}}^\bullet(X), d_W) \in \mathcal{C}^n$$

induces a map $\Omega_n^{\text{Witt}} \to \Omega_n^{SD}$. Note that if Y^{n+1} is a Witt space cobordism, then we have a corresponding cobordism $(Y, \mathbf{IC}_{\bar{m}}^\bullet(\operatorname{int} Y), \delta_W) \in Cob^{n+1}$, $\delta_W : \mathcal{D}\mathbf{IC}_{\bar{m}}^\bullet(\operatorname{int} Y)[n+1] \xrightarrow{\cong} \mathbf{IC}_{\bar{m}}^\bullet(\operatorname{int} Y)$.
We shall need the following lemma:

LEMMA 4.3. *Let Y^{n+1} be a compact pseudomanifold with boundary, $\boldsymbol{B}^\bullet \in D^b(Y)$ be constructible up to and including the boundary and $i : \operatorname{int} Y \hookrightarrow Y$ the inclusion. Then the canonical adjunction*

$$\boldsymbol{B}^\bullet \to Ri_* i^* \boldsymbol{B}^\bullet$$

is an isomorphism.

PROOF. Let $y \in \partial Y$. It suffices to prove that

$$\mathbf{H}^s(\mathbf{B}^\bullet)_y \longrightarrow \mathbf{H}^s(Ri_* i^* \mathbf{B}^\bullet)_y$$

is an isomorphism for all s, since the map is clearly an isomorphism over $\operatorname{int} Y$. But this follows from [**B$^+$84**, V, 10.22(4)], using a collar neighborhood of the boundary. □

Caveat. This lemma must not be applied to sheaves like $Ri_!\mathbf{B}^\bullet$, $\mathbf{B}^\bullet \in D^b(\operatorname{int} Y)$, or the dualizing complex $\mathbb{D}_Y^\bullet$, as these sheaves are not constructible up to and including the boundary. An application of the lemma to $Ri_!\mathbf{B}^\bullet$, for example, would yield the incorrect result $Ri_!\mathbf{B}^\bullet \cong Ri_* i^* Ri_! \mathbf{B}^\bullet \cong Ri_* \mathbf{B}^\bullet$.

The following lemma implies the claim:

LEMMA 4.4. *Let Y^{n+1} be a Witt space cobordism. Then*

$$\partial(Y, \boldsymbol{IC}^\bullet_{\bar{m}}(\operatorname{int} Y), \delta_W) \cong (\partial Y, \boldsymbol{IC}^\bullet_{\bar{m}}(\partial Y), d_W)$$

PROOF. Let Z be the space obtained from Y by attaching an open collar to the boundary:

$$Z = Y \cup_{\partial Y} \partial Y \times [0, 1)$$

Let $\bar{\jmath} : Y \hookrightarrow Z$ be the inclusion. By lemma 4.3,

$$Ri_* i^* \bar{\jmath}^* \mathbf{IC}^\bullet_{\bar{m}}(Z) \cong \bar{\jmath}^* \mathbf{IC}^\bullet_{\bar{m}}(Z).$$

The inclusion $\bar{\jmath} \circ j : \partial Y \hookrightarrow Z$ is normally nonsingular, so by theorem 5.4.1 [**GM83**],

$$(\bar{\jmath} \circ j)^* \mathbf{IC}^\bullet_{\bar{m}}(Z) \cong \mathbf{IC}^\bullet_{\bar{m}}(\partial Y)[1]$$

Using $\mathbf{IC}^\bullet_{\bar{m}}(\operatorname{int} Y) \cong i^* \bar{\jmath}^* \mathbf{IC}^\bullet_{\bar{m}}(Z)$ and lemma 4.1, we compute

$$\begin{aligned} j^! Ri_! \mathbf{IC}^\bullet_{\bar{m}}(\operatorname{int} Y) &\cong j^* Ri_* \mathbf{IC}^\bullet_{\bar{m}}(\operatorname{int} Y)[-1] \\ &\cong j^* Ri_* i^* \bar{\jmath}^* \mathbf{IC}^\bullet_{\bar{m}}(Z)[-1] \\ &\cong j^* \bar{\jmath}^* \mathbf{IC}^\bullet_{\bar{m}}(Z)[-1] \\ &\cong \mathbf{IC}^\bullet_{\bar{m}}(\partial Y), \end{aligned}$$

and $\partial \delta_W = d_W$ by uniqueness. □

It follows that with Ω_*^{ev} denoting the cobordism group associated to pseudo-manifolds having only even codimension strata (as introduced in [**GM80**]), one has the commutative diagram

$$\begin{array}{ccccc} \Omega_*^{\text{ev}} & \longrightarrow & \Omega_*^{\text{Witt}} & \longrightarrow & \Omega_*^{SD} \\ & \searrow^{\sigma} & \downarrow{\scriptstyle\sigma} & \swarrow_{\sigma} & \\ & & \mathbb{Z} & & \end{array}$$

CHAPTER 5

Lagrangian Structures and Ordered Resolutions

1. Statement of Result

This chapter provides a proof of the following result:

THEOREM 5.1. *Let X^n be a PL-variety which has an ordered resolution*

$$\{X, \dots, M^n\},$$

where M is a closed oriented PL-manifold. Then there exist a complex of sheaves $\boldsymbol{IC}^{\bullet}_{\mathcal{L}}(X) \in D^b(X)$, morphisms

$$\boldsymbol{IC}^{\bullet}_{\bar{m}}(X) \longrightarrow \boldsymbol{IC}^{\bullet}_{\mathcal{L}}(X) \longrightarrow \boldsymbol{IC}^{\bullet}_{\bar{n}}(X)$$

which factor the canonical morphism $\boldsymbol{IC}^{\bullet}_{\bar{m}}(X) \to \boldsymbol{IC}^{\bullet}_{\bar{n}}(X)$, and a self-duality isomorphism

$$\mathcal{D}\boldsymbol{IC}^{\bullet}_{\mathcal{L}}(X)[n] \cong \boldsymbol{IC}^{\bullet}_{\mathcal{L}}(X)$$

such that the following diagram commutes:

$$\begin{array}{ccc}
\boldsymbol{IC}^{\bullet}_{\mathcal{L}}(X) & \longrightarrow & \boldsymbol{IC}^{\bullet}_{\bar{n}}(X) \\
\uparrow \cong & & \uparrow \cong \\
\mathcal{D}\boldsymbol{IC}^{\bullet}_{\mathcal{L}}(X)[n] & \longrightarrow & \mathcal{D}\boldsymbol{IC}^{\bullet}_{\bar{m}}(X)[n]
\end{array}$$

By the algebraic framework developed in chapter 2, this is equivalent to the statement that a variety with an ordered resolution has Lagrangian structures along every odd-codimensional pure stratum. The proof is inductive, starting out on the manifold and following the resolution sequence. We apply both the algebraic machinery from chapter 2 as well as our work on the cobordism group Ω_*^{SD}. In general, the pushforward under the resolution map of the intersection chain complex on the manifold is not quasi-isomorphic to the complex $\mathbf{IC}^{\bullet}_{\mathcal{L}}$ constructed in the theorem. The pushforward will be self-dual since the map is proper, but it will not necessarily satisfy the required stalk conditions. Ordered resolutions are not assumed to be small maps.

2. The inductive set-up

Assume we are given an oriented closed manifold M^n and a sequence

$$\{X^{(m+1)}, X^{(m)}, \dots, X^{(1)}, X^{(0)} = M\}$$

of spaces constructed as in chapter 3. We shall review this construction here:

Associated to every $X^{(s)}$, $s = 0, \dots m$, we have a good embedding $E_s \hookrightarrow X^{(s)}_{b(s-1)}$ (recall that $X^{(s)}_{b(s-1)}$ is the bottom stratum of $X^{(s)}$, for the definition of "good" see section 2, chapter 3) a block fibration $F_s \to E_s \xrightarrow{\pi_s} B^{b(s)}$ and the block fibrations, constructed in chapter 3,

$$E(\zeta_s) \xrightarrow{p_s} E_s$$

and

$$E(\zeta_s^r) \xrightarrow{p_s^r} E_s,$$

where $E(\zeta_s)$ describes a regular neighborhood of E_s in $X^{(s)}$, and $E(\zeta_o^r)$ describes its boundary, that is $\partial E(\zeta_s) = E(\zeta_s^r)$. Then

$$X^{(s+1)} = (X^{(s)} - \mathrm{int}E(\zeta_s)) \cup_{E(\zeta_s^r)} M^{(s)},$$

where $M^{(s)}$ is the mapping cylinder of $\pi_s p_s^r : E(\zeta_s^r) \to B^{b(s)}$. The link of the bottom stratum $X^{(m+1)}_{b(m)} \cong B^{b(m)}$ in $X^{(m+1)}$ is the subvariety

$$V(x) = (p_m^r)^{-1} \pi_m^{-1}(x),$$

for $x \in B^{b(m)}$. Let $j_{m+1} : X^{(m+1)}_{b(m)} \hookrightarrow X^{(m+1)}$ be the closed inclusion and put $\Sigma = X^{(m+1)}_{b(m)}$.

Let $N^{(s)}$ be an open regular neighborhood of $E(\zeta_s^r)$ in $E(\zeta_s)$. By the collar neighborhood theorem for PL spaces, there exists a stratum preserving homeomorphism

$$M^{(s)} - B^{b(s)} = \partial E(\zeta_s) \times [0, 1) \xrightarrow{\cong} N^{(s)},$$

which extends by the identity to a homeomorphism

$$\phi_{s+1} : X^{(s+1)} - X^{(s+1)}_{b(s)} \xrightarrow{\cong} (X^{(s)} - \mathrm{int}E(\zeta_s)) \cup N^{(s)}.$$

We have inclusions

$$\iota_{s+1} : (X^{(s)} - \mathrm{int}E(\zeta_s)) \cup N^{(s)} \hookrightarrow X^{(s)}$$

and

$$i_{s+1} : X^{(s+1)} - X^{(s+1)}_{b(s)} \hookrightarrow X^{(s+1)}.$$

Set

$$\mathbf{L}^\bullet(X^{(0)}) = \mathbb{R}_M[n].$$

(instead of $\mathbf{IC}^\bullet_{\mathcal{L}}(X^{(s)})$, we shall use the shorter notation $\mathbf{L}^\bullet(X^{(s)})$.) The orientation of M provides an isomorphism $\mathbb{D}^\bullet_M \cong \mathbb{R}_M[n]$. Hence,

$$\mathcal{D}\mathbf{L}^\bullet(X^{(0)})[n] = \mathcal{D}(\mathbb{R}_M[n])[n] \cong \mathcal{D}\mathbb{R}_M \cong \mathbb{D}^\bullet_M \cong \mathbb{R}_M[n] = \mathbf{L}^\bullet(X^{(0)}),$$

which fixes the duality isomorphism $\mathcal{D}\mathbf{L}^\bullet(X^{(0)})[n] \cong \mathbf{L}^\bullet(X^{(0)})$, so that $\mathbf{L}^\bullet(X^{(0)})$ defines an element $\mathbf{L}^\bullet(X^{(0)}) \in SD(X^{(0)})$.

Now assume as an induction hypothesis that

$$\mathbf{L}^\bullet(X^{(m)}) \in SD(X^{(m)})$$

has been constructed. Define $k(s) = \bar{n}(n - b(s-1)) - n$, and $k'(s) = \bar{m}(n - b(s-1)) - n$, $s = 1, \dots, m+1$. The goal is to construct $\mathbf{L}^\bullet(X^{(m+1)}) \in SD(X^{(m+1)})$. We assume that $n - b(m)$ is odd, i.e. Σ is of odd codimension in $X^{(m+1)}$ (if $n - b(m)$ is even, then by theorem 2.10, no Lagrangian structure has to be chosen). Note that in this notation, the lower and upper middle perversity intersection chain sheaves on $X^{(m+1)}$ can be constructed as

$$\begin{aligned}\mathbf{IC}^{\bullet}_{\bar{m}}(X^{(m+1)}) &= \tau_{\leq k(m+1)-1} Ri_{m+1_*}\phi^*_{m+1}\iota^*_{m+1} \\ &\quad \tau_{\leq k'(m)} Ri_{m_*}\phi^*_m\iota^*_m \cdots \tau_{\leq k'(1)} Ri_{1_*}\phi^*_1\iota^*_1\mathbb{R}_M[n],\end{aligned}$$

and

$$\begin{aligned}\mathbf{IC}^{\bullet}_{\bar{n}}(X^{(m+1)}) &= \tau_{\leq k(m+1)} Ri_{m+1_*}\phi^*_{m+1}\iota^*_{m+1} \\ &\quad \tau_{\leq k(m)} Ri_{m_*}\phi^*_m\iota^*_m \cdots \tau_{\leq k(1)} Ri_{1_*}\phi^*_1\iota^*_1\mathbb{R}_M[n],\end{aligned}$$

Set

$$\mathbf{P}^{\bullet} = \phi^*_{m+1}\iota^*_{m+1}\mathbf{L}^{\bullet}(X^{(m)}).$$

As ι_{m+1} is an open inclusion and ϕ_{m+1} a homeomorphism,

$$\mathcal{D}\mathbf{L}^{\bullet}(X^{(m)})[n] \cong \mathbf{L}^{\bullet}(X^{(m)})$$

induces

$$\mathcal{D}\mathbf{P}^{\bullet}[n] \cong \mathbf{P}^{\bullet}.$$

Put

$$\begin{aligned}\mathbf{S}^{\bullet} &= Ri_{m+1_*}\mathbf{P}^{\bullet} \\ \mathbf{S}^{\bullet}_1 &= \tau_{\leq k(m+1)-1}\mathbf{S}^{\bullet} \\ \mathbf{S}^{\bullet}_2 &= \tau_{\leq k(m+1)}\mathbf{S}^{\bullet}\end{aligned}$$

By section 9 in [**GM83**], $\mathcal{D}\mathbf{P}^{\bullet}[n] \cong \mathbf{P}^{\bullet}$ induces $\mathcal{D}\mathbf{S}^{\bullet}_1[n] \cong \mathbf{S}^{\bullet}_2$.

3. Construction of a nonsingular pairing on $\mathbf{H}^k(j^*\mathbf{S}^{\bullet})$

Write $k = k(m+1), i = i_{m+1}$, and $j = j_{m+1}$. We have

$$\begin{aligned}\mathbf{H}^l(j^*\mathbf{S}^{\bullet}_1) &= 0, \quad \text{for} \quad l \geq k \\ \mathbf{H}^l(j^*\mathbf{S}^{\bullet}_2) &= 0, \quad \text{for} \quad l \geq k+1\end{aligned}$$

and using $\mathcal{D}\mathbf{S}^{\bullet}_1[n] \cong \mathbf{S}^{\bullet}_2$, we obtain

$$\begin{aligned}\mathbf{H}^l(j^!\mathbf{S}^{\bullet}_1) &= 0, \quad \text{for} \quad l \leq k \\ \mathbf{H}^l(j^!\mathbf{S}^{\bullet}_2) &= 0, \quad \text{for} \quad l \leq k+1.\end{aligned}$$

Consider the triangles

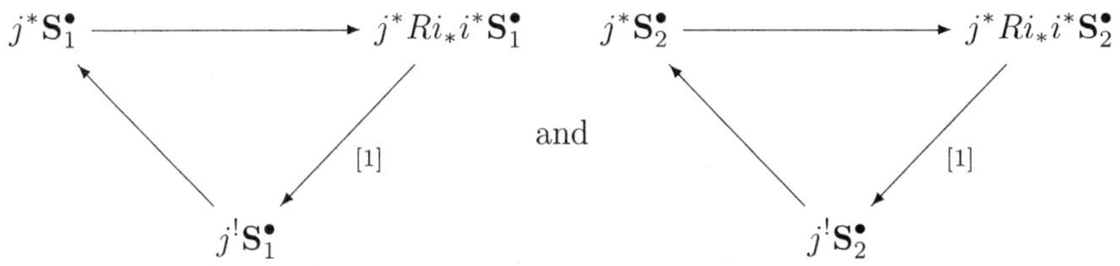

On cohomology they induce

$$\mathbf{H}^k(j^!\mathbf{S}^{\bullet}_1) \to \mathbf{H}^k(j^*\mathbf{S}^{\bullet}_1) \to \mathbf{H}^k(j^*Ri_*i^*\mathbf{S}^{\bullet}_1) \to \mathbf{H}^{k+1}(j^!\mathbf{S}^{\bullet}_1) \to \mathbf{H}^{k+1}(j^*\mathbf{S}^{\bullet}_1)$$

and

$$\mathbf{H}^k(j^!\mathbf{S}^{\bullet}_2) \to \mathbf{H}^k(j^*\mathbf{S}^{\bullet}_2) \to \mathbf{H}^k(j^*Ri_*i^*\mathbf{S}^{\bullet}_2) \to \mathbf{H}^{k+1}(j^!\mathbf{S}^{\bullet}_2) \to \mathbf{H}^{k+1}(j^*\mathbf{S}^{\bullet}_2)$$

From $\mathbf{H}^k(j^*\mathbf{S}_1^\bullet) = 0$, $\mathbf{H}^{k+1}(j^*\mathbf{S}_1^\bullet) = 0$, $\mathbf{H}^k(j^!\mathbf{S}_2^\bullet) = 0$, $\mathbf{H}^{k+1}(j^!\mathbf{S}_2^\bullet) = 0$ it follows that we have isomorphisms

$$\mathbf{H}^k(j^*Ri_*i^*\mathbf{S}_1^\bullet) \xrightarrow{\cong} \mathbf{H}^{k+1}(j^!\mathbf{S}_1^\bullet)$$

and

$$\mathbf{H}^k(j^*\mathbf{S}_2^\bullet) \xrightarrow{\cong} \mathbf{H}^k(j^*Ri_*i^*\mathbf{S}_2^\bullet).$$

As $\tau_{\leq k(m)}\mathbf{L}^\bullet(X^{(m)}) \to \mathbf{L}^\bullet(X^{(m)})$ is a quasi-isomorphism, the canonical maps $\mathbf{S}_1^\bullet \to \mathbf{S}_2^\bullet \to \mathbf{S}^\bullet$ induce quasi-isomorphisms over $X^{(m+1)}-\Sigma$, so we have quasi-isomorphisms

$$Ri_*i^*\mathbf{S}_1^\bullet \xrightarrow{\cong} Ri_*i^*\mathbf{S}_2^\bullet \xrightarrow{\cong} Ri_*i^*\mathbf{S}^\bullet \simeq \mathbf{S}^\bullet,$$

and isomorphisms

$$\mathbf{H}^k(j^*Ri_*i^*\mathbf{S}_1^\bullet) \xrightarrow{\cong} \mathbf{H}^k(j^*Ri_*i^*\mathbf{S}_2^\bullet) \xrightarrow{\cong} \mathbf{H}^k(j^*\mathbf{S}^\bullet).$$

Summarizing, we have constructed canonical isomorphisms

$$\begin{array}{ccccc} & & \mathbf{H}^k(j^*\mathbf{S}^\bullet) & & \\ & & \uparrow \cong & & \\ \mathbf{H}^k(j^*\mathbf{S}_2^\bullet) & \xrightarrow{\cong} & \mathbf{H}^k(j^*Ri_*i^*\mathbf{S}_2^\bullet) & & \\ & & \cong\uparrow & & \\ & & \mathbf{H}^k(j^*Ri_*i^*\mathbf{S}_1^\bullet) & \xrightarrow{\cong} & \mathbf{H}^{k+1}(j^!\mathbf{S}_1^\bullet) \end{array}$$

We proceed to construct the pairing on $\mathbf{H}^k(j^*\mathbf{S}^\bullet)$:
The duality isomorphism $\mathcal{D}\mathbf{S}_1^\bullet[n] \cong \mathbf{S}_2^\bullet$ induces a nonsingular pairing on derived sheaves

$$\mathbf{H}^k(j^*\mathbf{S}_2^\bullet) \otimes \mathbf{H}^{k+1}(j^!\mathbf{S}_1^\bullet) \longrightarrow \mathbb{R}_\Sigma.$$

Under the above identifications, this induces a nonsingular pairing

$$\mathbf{H}^k(j^*\mathbf{S}^\bullet) \otimes \mathbf{H}^k(j^*\mathbf{S}^\bullet) \longrightarrow \mathbb{R}_\Sigma,$$

so that the following diagram commutes:

$$\begin{array}{ccc} \mathbf{H}^k(j^*\mathbf{S}_2^\bullet) & \xrightarrow{\cong} & \mathbf{H}^k(j^*\mathbf{S}^\bullet) \\ \otimes & & \otimes \\ \mathbf{H}^{k+1}(j^!\mathbf{S}_1^\bullet) & \xrightarrow{\cong} & \mathbf{H}^k(j^*\mathbf{S}^\bullet) \\ \downarrow & & \downarrow \\ \mathbb{R}_\Sigma & & \mathbb{R}_\Sigma \end{array}$$

4. Stalks of $\mathbf{H}^k(j^*\mathbf{S}^\bullet)$ as the hypercohomology of the link of Σ

4.1. The Leray-sheaf of the fibration $\pi_m p_m^r$. For $U \subset \Sigma$ open, let

$$V(U) = (p_m^r)^{-1}\pi_m^{-1}(U).$$

We may regard $V(U)$ either as a subspace of $X^{(m)}$ or as a subspace of $X^{(m+1)}$. Note $\mathbf{S}^\bullet|_{V(U)} \cong \mathbf{P}^\bullet|_{V(U)} \cong \mathbf{L}^\bullet(X^{(m)})|_{V(U)}$. We define the Leray-sheaf of $\pi_m p_m^r$ as

$$\mathbf{G} = \mathrm{Sheaf}(U \mapsto \mathcal{H}^k(V(U); \mathbf{L}^\bullet(X^{(m)}))).$$

For $x \in \Sigma$, there is a canonical morphism $\mathbf{G}_x \to \mathcal{H}^k(V(x); \mathbf{L}^\bullet(X^{(m)}))$. As $\pi_m p_m^r$ is a proper map between locally compact spaces, this morphism is in fact an isomorphism:

$$\mathbf{G}_x \cong \mathcal{H}^k(V(x); \mathbf{L}^\bullet(X^{(m)})),$$

see [**Ive86**, theorem III.6.2].

4.2. The Leray-sheaf is isomorphic to $\mathbf{H}^k(j^*\mathbf{S}^\bullet)$. We need the following Vietoris-Begle theorem:

LEMMA 5.1. *Suppose X is a PL-variety, $\boldsymbol{B}^\bullet \in D^b(X)$, and $\pi : \mathbb{R}^m \times X \to X$ the projection. Then adjunction is a quasi-isomorphism $\boldsymbol{B}^\bullet \xrightarrow{\cong} R\pi_*\pi^*\boldsymbol{B}^\bullet$.*

A proof can be found in [**B$^+$84**, V, 3.13]. The next fact is contained in the proof of lemma 10.14 in [**B$^+$84**, V].

LEMMA 5.2. *Let X be a PL-variety, $\boldsymbol{A}^\bullet \in D^b(\mathbb{R}^m \times X)$ and $\{0\} \times X \xrightarrow{j} \mathbb{R}^m \times X \xrightarrow{\pi} \{0\} \times X$. Then canonically*

$$\boldsymbol{A}^\bullet \cong \pi^* j^* \boldsymbol{A}^\bullet.$$

The following proposition identifies the Leray sheaf of $\pi_m p_m^r$ with the cohomology sheaf of the restriction of $\mathbf{S}^\bullet$ to Σ.

PROPOSITION 5.1. *$\boldsymbol{H}^k(j^*\boldsymbol{S}^\bullet) \cong \boldsymbol{G}$.*

PROOF. $M^{(m)}$ is the mapping cylinder of $\pi_m p_m^r : E(\zeta_m^r) \to \Sigma$. It comes with a retraction $\rho : \mathrm{int}M^{(m)} \to \Sigma$. Using a collar, we may assume that $V(\Sigma)$ is pushed off the boundary of $M^{(m)}$ so that $V(U) \subset \mathrm{int}M^{(m)}$ for all $U \subset \Sigma$ open. Let $\{U\}$ be a neighborhood basis for Σ such that there exist isomorphisms $\rho^{-1}(U) \cong \mathbb{R}^{b(m)} \times c^\circ V$ which restrict as follows:

$$\begin{array}{ccc}
\rho^{-1}(U) & \xrightarrow{\cong} & \mathbb{R}^{b(m)} \times c^\circ V \\
\uparrow & & \uparrow \\
V(U) & \xrightarrow{\cong} & \mathbb{R}^{b(m)} \times \{\frac{1}{2}\} \times V,
\end{array}$$

where $\{\frac{1}{2}\} \times V \hookrightarrow [0,1) \times V/_{\{0\}\times V} = c^\circ V$. Moreover, $\rho^{-1}(U) - \Sigma \xrightarrow{\cong} \mathbb{R}^{b(m)} \times (0,1) \times V$ under the above isomorphism. Define the Leray-sheaf of ρ as

$$\mathbf{A} = \mathrm{Sheaf}(U \mapsto \mathcal{H}^k(\rho^{-1}(U); \mathbf{S}^\bullet)).$$

Now $\mathbf{H}^k(j^*\mathbf{S}^\bullet) = \mathrm{Sheaf}(U \mapsto \mathcal{H}^k(U; j^*\mathbf{S}^\bullet))$, so restriction of sections

$$\mathcal{H}^k(\rho^{-1}(U); \mathbf{S}^\bullet) \to \mathcal{H}^k(U; \mathbf{S}^\bullet)$$

induces a map $\mathbf{A} \to \mathbf{H}^k(j^*\mathbf{S}^\bullet)$. If $x \in \Sigma$ and U is a small neighborhood of x in Σ, then $\mathbf{H}^k(j^*\mathbf{S}^\bullet)_x \cong \mathcal{H}^k(\rho^{-1}(U); \mathbf{S}^\bullet)$, as $\rho^{-1}(U)$ is a distinguished neighborhood of x, see [**GM83**]. So in fact $\mathbf{A} \xrightarrow{\cong} \mathbf{H}^k(j^*\mathbf{S}^\bullet)$.
Restriction of sections $\mathcal{H}^k(\rho^{-1}(U); \mathbf{S}^\bullet) \to \mathcal{H}^k(V(U); \mathbf{S}^\bullet)$ defines $\mathbf{A} \to \mathbf{G}$. We calculate

$$\mathcal{H}^k(\rho^{-1}(U); \mathbf{S}^\bullet) = \mathcal{H}^k(\rho^{-1}(U); Ri_*\mathbf{P}^\bullet) \cong \mathcal{H}^k(\rho^{-1}(U) - \Sigma; \mathbf{P}^\bullet).$$

Let $Y = \mathbb{R}^{b(m)} \times V$, so that $\rho^{-1}(U) - \Sigma \cong (0,1) \times Y$. With $\{\frac{1}{2}\} \times Y \overset{\bar{j}}{\hookrightarrow} (0,1) \times Y \overset{\pi}{\longrightarrow} \{\frac{1}{2}\} \times Y$ we compute

$$\mathcal{H}^k((0,1) \times Y; \mathbf{P}^\bullet) \cong \mathcal{H}^k((0,1) \times Y; \pi^* \bar{j}^* \mathbf{P}^\bullet),$$

using lemma 5.2, and

$$\mathcal{H}^k((0,1) \times Y; \pi^* \bar{j}^* \mathbf{P}^\bullet) \cong \mathcal{H}^k(\{\tfrac{1}{2}\} \times Y; R\pi_* \pi^* \bar{j}^* \mathbf{P}^\bullet) \cong \mathcal{H}^k(\{\tfrac{1}{2}\} \times Y; \bar{j}^* \mathbf{P}^\bullet),$$

using lemma 5.1. We conclude that $\mathcal{H}^k(\rho^{-1}(U) - \Sigma; \mathbf{P}^\bullet) \cong \mathcal{H}^k(V(U); \mathbf{P}^\bullet)$ and therefore $\mathbf{A} \overset{\cong}{\longrightarrow} \mathbf{G}$. So via $\mathbf{A}$, we identify

$$\mathbf{H}^k(j^* \mathbf{S}^\bullet) \overset{\cong}{\longleftarrow} \mathbf{A} \overset{\cong}{\longrightarrow} \mathbf{G}.$$

□

In particular, it follows that for $x \in \Sigma$,

$$\mathbf{H}^k(j^* \mathbf{S}^\bullet)_x \cong \mathcal{H}^k(V(x); \mathbf{L}^\bullet(X^{(m)})).$$

5. The restriction of $\mathbf{L}^\bullet(X^{(m)})$ to $V(x)$ is self-dual

Let T be an open regular neighborhood of $V(x)$ in $X^{(m+1)}$. The point x has a distinguished neighborhood N in $X^{(m+1)}$ such that $N \cong \operatorname{int} D^{b(m)} \times c^\circ(V)$. Hence

$$T = \operatorname{nbhd}(V, X^{(m+1)}) \cong \operatorname{nbhd}(V, N) \cong$$

$$\operatorname{nbhd}(V, \operatorname{int} D^{b(m)} \times c^\circ(V)) \cong \operatorname{nbhd}(V, \mathbb{R}^{b(m)} \times (0,1) \times V) \cong$$

$$\operatorname{nbhd}(V, \mathbb{R}^{b(m)+1} \times V) \cong \mathbb{R}^{b(m)+1} \times V.$$

So the tubular neighborhood T is isomorphic to the total space of a trivial vector bundle over V. Hence, the inclusion $\bar{j} : V \hookrightarrow X^{(m+1)}$ is normally nonsingular in the sense of [**GM83**, section 5.4.2]. Thus the restriction of $\mathbf{P}^\bullet$ to V is self-dual as can be seen from

$$\mathcal{D}(\bar{j}^* \mathbf{P}^\bullet)[n + b(m) + 1] \cong \bar{j}^! \mathcal{D}\mathbf{P}^\bullet[n][b(m)+1] \cong \bar{j}^! \mathbf{P}^\bullet[b(m)+1] \cong \bar{j}^* \mathbf{P}^\bullet.$$

It is convenient to set

$$\mathbf{L}^\bullet(V) = \mathbf{P}^\bullet|_V[-b(m) - 1].$$

In this notation,

$$\boxed{\mathcal{D}\mathbf{L}^\bullet(V)[\dim V] \cong \mathbf{L}^\bullet(V),}$$

so that $\mathbf{L}^\bullet(V)$ is self-dual. Note that $\dim V = n - b(m) - 1$ is even, say $\dim V = 2v$. So $k = k(m+1) = \frac{1}{2}(n - b(m) - 1) - n = v - n$ and $k + b(m) + 1 = -v$, from which we see that

$$\mathcal{H}^k(V(x); \mathbf{L}^\bullet(X^{(m)})) \cong \mathcal{H}^k(V; \mathbf{P}^\bullet|_V) = \mathcal{H}^k(V; \mathbf{L}^\bullet(V)[b(m)+1]) =$$

$$\mathcal{H}^{k+b(m)+1}(V; \mathbf{L}^\bullet(V)) = \mathcal{H}^{-v}(V; \mathbf{L}^\bullet(V)).$$

Hence

$$\boxed{\mathbf{H}^k(j^* \mathbf{S}^\bullet)_x \cong \mathcal{H}^{-v}(V; \mathbf{L}^\bullet(V))}$$

(using section 4) and there is a commutative diagram

$$\begin{array}{ccc} \mathbf{H}^k(j^*\mathbf{S}^\bullet)_x & \xrightarrow{\cong} & \mathcal{H}^{-v}(V(x);\mathbf{L}^\bullet(V)) \\ \otimes & & \otimes \\ \mathbf{H}^k(j^*\mathbf{S}^\bullet)_x & \xrightarrow{\cong} & \mathcal{H}^{-v}(V(x);\mathbf{L}^\bullet(V)) \\ \downarrow & & \downarrow \\ \mathbb{R} & & \mathbb{R} \end{array}$$

where the pairing on the hypercohomology groups is induced by $\mathcal{D}\mathbf{L}^\bullet(V)[2v] \cong \mathbf{L}^\bullet(V)$.

6. The construction of a Lagrangian subsheaf of $\mathbf{H}^k(j^*\mathbf{S}^\bullet)$

6.1. The Leray-sheaf of the fibration $\pi_m p_m$. Recall that the block fibration

$$E(\zeta_m) \xrightarrow{p_m} E_m$$

describes a regular neighborhood of E_m in $X^{(m)}$, see section 2, or chapter 3. For $U \subset \Sigma \cong B^{b(m)}$ open, let $W(U) = p_m^{-1}\pi_m^{-1}(U)$, and for $x \in \Sigma$ let $W(x) = p_m^{-1}\pi_m^{-1}(x)$. Then $W(U)$ and $W(x)$ are subvarieties of $X^{(m)}$ (but not of $X^{(m+1)}$), $W(x)$ is compact and

$$\boxed{\partial W(x) = V(x).}$$

Write $w = \dim W(x)$. We define the Leray-sheaf of $\pi_m p_m$:

$$\mathbf{E} = \mathrm{Sheaf}(U \mapsto \mathcal{H}^k(W(U);\mathbf{L}^\bullet(X^{(m)}))).$$

Analogously to section 4.1, as $\pi_m p_m$ is a proper map,

$$\mathbf{E}_x \xrightarrow{\cong} \mathcal{H}^k(W(x);\mathbf{L}^\bullet(X^{(m)})).$$

We have inclusion maps

$$\begin{array}{ccccc} \mathrm{int}W(\cdot) & \overset{\bar{\imath}}{\hookrightarrow} & W(\cdot) & \subset X^{(m)} \\ & & \uparrow \bar{\jmath} & \\ & & V(\cdot) & \end{array}$$

where the dot signifies either a point $x \in \Sigma$ or an open $U \subset \Sigma$.

Restriction of sections induces a map of presheaves

$$(U \mapsto \mathcal{H}^k(W(U);\mathbf{L}^\bullet(X^{(m)}))) \longrightarrow (U \mapsto \mathcal{H}^k(V(U);\mathbf{L}^\bullet(X^{(m)})))$$

and so a map $\mathbf{E} \to \mathbf{G} \cong \mathbf{H}^k(j^*\mathbf{S}^\bullet)$. Let

$$\boxed{\bar{\mathbf{E}} = \mathbf{Im}(\mathbf{E} \to \mathbf{H}^k(j^*\mathbf{S}^\bullet)),}$$

a subsheaf of $\mathbf{H}^k(j^*\mathbf{S}^\bullet)$.

The remainder of section 6 is devoted to showing that $\bar{\mathbf{E}}$ is a Lagrangian subsheaf of $\mathbf{H}^k(j^*\mathbf{S}^\bullet)$. We will use the results of chapter 4.

6.2. The restriction of $\mathbf{L}^\bullet(X^{(m)})$ to $\operatorname{int} W(x)$ is self-dual. An open regular neighborhood of $\operatorname{int}W(x)$ in $X^{(m)}$ is isomorphic to the interior of the block $\gamma(\Delta) \in \gamma$, with $\Delta \in B(\alpha)$, $x \in \Delta$, $\dim \Delta = b(m)$, where we use the notation of chapter 3. So $\operatorname{int}W \hookrightarrow X^{(m)}$ is a normally nonsingular embedding and, as in section 5,

$$\mathcal{D}(\bar{\imath}^*\mathbf{L}^\bullet(X^{(m)})|_{W(x)})[n + b(m)] \cong \bar{\imath}^*\mathbf{L}^\bullet(X^{(m)})|_{W(x)}.$$

Set

$$\mathbf{L}^\bullet(W) = \mathbf{L}^\bullet(X^{(m)})|_{W(x)}[-b(m)].$$

Then

$$\boxed{\mathcal{D}(\bar{\imath}^*\mathbf{L}^\bullet(W))[w] \cong \bar{\imath}^*\mathbf{L}^\bullet(W),}$$

so that $\bar{\imath}^*\mathbf{L}^\bullet(W)$ is self-dual.

6.3. $\bar{\mathrm{E}}$ is a Lagrangian subsheaf of $\mathbf{H}^k(j^*\mathbf{S}^\bullet)$.

THEOREM 5.2. *Let $\mathcal{H}^k(W(x); \boldsymbol{L}^\bullet(X^{(m)})) \to \mathcal{H}^k(V(x); \boldsymbol{L}^\bullet(X^{(m)}))$ be the map induced by restriction of sections. The image of this map is a Lagrangian subspace in*

$$\mathcal{H}^k(V(x); \boldsymbol{L}^\bullet(X^{(m)})) = \mathcal{H}^{-v}(V(x); \boldsymbol{L}^\bullet(V)).$$

PROOF. Put $\mathbf{B}^\bullet = \bar{\imath}^*\mathbf{L}^\bullet(W)$. From section 6.2 we have the self-duality isomorphism $\delta : \mathcal{D}\mathbf{B}^\bullet[w] \cong \mathbf{B}^\bullet$. By lemma 4.3,

$$R\bar{\imath}_*\mathbf{B}^\bullet \cong \mathbf{L}^\bullet(W).$$

We calculate

$$\bar{\jmath}^*\mathbf{L}^\bullet(W) = \bar{\jmath}^*\mathbf{L}^\bullet(X^{(m)})|_W[-b(m)] = \bar{\jmath}^*\mathbf{L}^\bullet(X^{(m)})|_W[-b(m) - 1][1] =$$

$$\mathbf{L}^\bullet(X^{(m)})|_V[-b(m) - 1][1] = \mathbf{L}^\bullet(V)[1]$$

and conclude

$$\bar{\jmath}^! R\bar{\imath}_!\mathbf{B}^\bullet \cong \bar{\jmath}^* R\bar{\imath}_*\mathbf{B}^\bullet[-1] \cong \mathbf{L}^\bullet(V),$$

using lemma 4.1. Note that this means that $(W(x), \mathbf{B}^\bullet, \delta) \in Cob^w$ defines an admissible cobordism whose boundary is

$$\partial(W(x), \mathbf{B}^\bullet, \delta) = (V(X), \mathbf{L}^\bullet(V), \partial\delta).$$

With $\mathbf{E}^\bullet = R\bar{\imath}_!\mathbf{B}^\bullet$, we apply theorem 4.1. It follows that

$$\operatorname{im}(\mathcal{H}^{-v-1}(W; R\bar{\imath}_*\bar{\imath}^*\mathbf{E}^\bullet) \to \mathcal{H}^{-v}(W; \bar{\jmath}_*\bar{\jmath}^!\mathbf{E}^\bullet))$$

is a Lagrangian subspace in $\mathcal{H}^{-v}(V; \bar{\jmath}^!\mathbf{E}^\bullet)$. The following calculations identify the above two hypercohomology groups as the groups mentioned in the statement of the theorem:

$$\begin{aligned}
\mathcal{H}^{-v-1}(W; R\bar{\imath}_*\bar{\imath}^*\mathbf{E}^\bullet) &\cong \mathcal{H}^{-v-1}(W; R\bar{\imath}_*\bar{\imath}^* R\bar{\imath}_!\bar{\imath}^*\mathbf{L}^\bullet(X^{(m)})|[-b(m)]) \\
&\cong \mathcal{H}^{-v-1-b(m)}(W; R\bar{\imath}_*\bar{\imath}^*\mathbf{L}^\bullet(X^{(m)})|) \\
&\cong \mathcal{H}^{-v-1-b(m)}(W; \mathbf{L}^\bullet(X^{(m)})) \\
&\cong \mathcal{H}^k(W; \mathbf{L}^\bullet(X^{(m)}))
\end{aligned}$$

and

$$\begin{aligned}\mathcal{H}^{-v}(W;\bar{\jmath}_*\bar{\jmath}^!\mathbf{E}^\bullet) &\cong \mathcal{H}^{-v}(V;\bar{\jmath}^! R\bar{\imath}_!\bar{\imath}^*\mathbf{L}^\bullet(X^{(m)})[-b(m)]) \\ &\cong \mathcal{H}^{-v-b(m)}(V;\bar{\jmath}^* R\bar{\imath}_*\bar{\imath}^*\mathbf{L}^\bullet(X^{(m)})[-1]) \\ &\cong \mathcal{H}^{-v-b(m)-1}(V;\bar{\jmath}^*\mathbf{L}^\bullet(X^{(m)})) \\ &\cong \mathcal{H}^{k}(V;\mathbf{L}^\bullet(X^{(m)})),\end{aligned}$$

using both lemma 4.1 and lemma 4.3, as well as $k+b(m)+1=-v$. □

The theorem shows that $\mathrm{im}(\mathbf{E}_x \to \mathbf{G}_x)$ is a Lagrangian subspace in $\mathbf{G}_x \cong \mathbf{H}^k(j^*\mathbf{S}^\bullet)_x$. But $\mathrm{im}(\mathbf{E}_x \to \mathbf{H}^k(j^*\mathbf{S}^\bullet)_x) = \bar{\mathbf{E}}_x$. So indeed $\bar{\mathbf{E}} \subset \mathbf{H}^k(j^*\mathbf{S}^\bullet)$ is a Lagrangian subsheaf.

7. The definition of $\mathbf{L}^\bullet(X^{(m+1)})$

The stalk conditions of $\mathbf{L}^\bullet(X^{(m)})$ imply $\mathbf{H}^k(\mathbf{S}^\bullet)_x = 0$ for $x \notin \Sigma$. This observation shows that if we extend $\bar{\mathbf{E}} \subset \mathbf{H}^k(j^*\mathbf{S}^\bullet)$ by zero outside of Σ, we obtain a subsheaf of $\mathbf{H}^k(\mathbf{S}^\bullet)$, which we shall again denote by $\bar{\mathbf{E}}$:

$$\bar{\mathbf{E}} \subset \mathbf{H}^k(\mathbf{S}^\bullet).$$

Let $\mathcal{O}(\mathbf{P}^\bullet)$ be the lifting obstruction of $\mathbf{P}^\bullet$, and $\mathcal{L} \to \mathcal{O}(\mathbf{P}^\bullet)$ be the Lagrangian structure given by the Lagrangian subsheaf $\bar{\mathbf{E}} \subset \mathbf{H}^k(j^*Ri_*\mathbf{P}^\bullet)$, as explained in remark 2.4. Define

$$\boxed{\mathbf{L}^\bullet(X^{(m+1)}) = \mathbf{P}^\bullet \boxplus \mathcal{L} \in SD(X^{(m+1)}).}$$

By theorem 2.2, there exist morphisms

$$\mathbf{IC}^\bullet_{\bar{m}}(X^{(m+1)}) \longrightarrow \mathbf{L}^\bullet(X^{(m+1)}) \longrightarrow \mathbf{IC}^\bullet_{\bar{n}}(X^{(m+1)})$$

factoring the canonical $\mathbf{IC}^\bullet_{\bar{m}}(X^{(m+1)}) \to \mathbf{IC}^\bullet_{\bar{n}}(X^{(m+1)})$, such that

$$\begin{array}{ccc} \mathbf{L}^\bullet(X^{(m+1)}) & \longrightarrow & \mathbf{IC}^\bullet_{\bar{n}}(X^{(m+1)}) \\ \uparrow \cong & & \uparrow \cong \\ \mathcal{D}\mathbf{L}^\bullet(X^{(m+1)})[n] & \longrightarrow & \mathcal{D}\mathbf{IC}^\bullet_{\bar{m}}(X^{(m+1)})[n] \end{array}$$

commutes. This finishes the inductive step and completes the proof of theorem 5.1.

APPENDIX A

On Signs

This appendix discusses in detail the signs of the various morphisms and pairings employed throughout the paper. In particular, we will answer the following question, which is of importance in applying the theory: Let X^n be a stratified pseudomanifold, $U_k = X - X_{n-k}$, $\mathbf{A}^\bullet \in SD(U_k)$ as in sections 1 and 2 of chapter 2. Lemma 2.3 asserts that the lifting obstruction $\mathcal{O}(\mathbf{A}^\bullet)$ (definition 2.1) inherits a self-duality isomorphism $\delta : \mathcal{D}\mathcal{O}(\mathbf{A}^\bullet)[n+1] \xrightarrow{\cong} \mathcal{O}(\mathbf{A}^\bullet)$ from the isomorphism $d : \mathcal{D}\mathbf{A}^\bullet[n] \to \mathbf{A}^\bullet$ associated to $\mathbf{A}^\bullet$. In remark 2.4, we point out that δ induces a nonsingular pairing δ' on $\mathcal{O}(\mathbf{A}^\bullet)$, regarded as a local coefficient system along the pure stratum $U_{k+1} - U_k$. It is for this pairing δ' that Lagrangian subspaces have to be chosen. Thus it is important to know whether δ' is symmetric or anti-symmetric. We will show that the answer is indeed as expected: δ' is symmetric when the dimension of the links of the components of $U_{k+1} - U_k$ is congruent 0 mod 4, and anti-symmetric when that dimension is congruent 2 mod 4.

In this paper, we follow all the standard sign conventions for differential complexes. Let us briefly summarize the facts that we need. Let A be the abelian category of vector spaces over a given field and let $C(A)$ denote the associated category of differential complexes. A morphism of complexes $\phi : X^\bullet \longrightarrow \mathrm{Hom}^\bullet(Y^\bullet, Z^\bullet)$, with $X^\bullet, Y^\bullet, Z^\bullet \in C(A)$, induces a pairing

$$p(\phi) : X^\bullet \otimes Y^\bullet \longrightarrow Z^\bullet$$

by setting

$$p^n(\phi)(x \otimes y) = \phi^p(x)_q(y),$$

where $p + q = n$, $x \in X^p$, $y \in Y^q$. Here $\phi^p(x) \in \prod_l \mathrm{Hom}(Y^l, Z^{l+p})$ and $\phi^p(x)_q$ denotes the q-entry of that sequence. Now fix $D^\bullet \in C(A)$ and put $DX^\bullet = \mathrm{Hom}^\bullet(X^\bullet, D^\bullet)$. Given a morphism $\phi : X^\bullet \longrightarrow Y^\bullet$, the induced morphism

$$D\phi = \mathrm{Hom}^\bullet(\phi, D^\bullet) : DY^\bullet \longrightarrow DX^\bullet$$

is given by

$$(D\phi)^n(f) = (f_l \circ \phi^l)_{l\in\mathbb{Z}},$$

where $f = (f_l)_{l\in\mathbb{Z}} \in \prod_l \mathrm{Hom}(Y^l, D^{l+n})$. In particular, for $Y^\bullet = DX^\bullet$ and $\phi : X^\bullet \longrightarrow DX^\bullet$ we get the morphism $D\phi : D(DX^\bullet) \longrightarrow DX^\bullet$. There is a canonical morphism

$$ev : X^\bullet \to DDX^\bullet$$

defined as follows: For $x \in X^n$, $ev^n(x)$ is an element of $\mathrm{Hom}^n(DX^\bullet, D^\bullet) = \prod_l \mathrm{Hom}((DX^\bullet)^l, D^{l+n})$ whose l-entry $ev^n(x)_l : \mathrm{Hom}^l(X^\bullet, D^\bullet) \longrightarrow D^{l+n}$ is given by

$$ev^n(x)_l(g) = (-1)^{nl} g_n(x),$$

where $g = (g_s)_{s\in\mathbb{Z}} \in \prod_s \operatorname{Hom}(X^s, D^{s+l})$. The identification

$$sn : \operatorname{Hom}^\bullet(X^\bullet[-1], Y^\bullet) \xrightarrow{\cong} \operatorname{Hom}^\bullet(X^\bullet, Y^\bullet)[1]$$

introduces sign changes in alternating degrees and is defined as

$$sn^n(f)_l = (-1)^{n+1} f_{l+1},$$

$f = (f_l) \in \prod_l \operatorname{Hom}(X^{l-1}, Y^{l+n}) = \operatorname{Hom}^n(X^\bullet[-1], Y^\bullet)$ (there is no sign change in the identification $\operatorname{Hom}^\bullet(X^\bullet, Y^\bullet[1]) \cong \operatorname{Hom}^\bullet(X^\bullet, Y^\bullet)[1]$). In particular, sn identifies

$$sn : D(X^\bullet[-1]) \xrightarrow{\cong} (DX^\bullet)[1]. \tag{28}$$

For the tensor product of two complexes, we have the "switch"-morphism

$$sw : X^\bullet \otimes Y^\bullet \longrightarrow Y^\bullet \otimes X^\bullet,$$

$$sw^n(x \otimes y) = (-1)^{pq} y \otimes x,$$

$p + q = n$, $x \in X^p$, $y \in Y^q$.

LEMMA A.1. *Let $\phi : X^\bullet \longrightarrow DX^\bullet$ be a morphism of complexes. Then*

$$\begin{array}{ccc} X^\bullet & \xrightarrow{\phi} & DX^\bullet \\ {\scriptstyle ev}\downarrow & \nearrow{\scriptstyle D\phi} & \\ DDX^\bullet & & \end{array}$$

commutes if and only if

$$\begin{array}{ccc} X^\bullet \otimes X^\bullet & \xrightarrow{p(\phi)} & D^\bullet \\ {\scriptstyle sw}\downarrow & \nearrow{\scriptstyle p(\phi)} & \\ X^\bullet \otimes X^\bullet & & \end{array}$$

commutes.

A few remarks on the sign convention in axiom (**SD4**) of definition 2.2 are in order. In (**SD4**), we require the self-duality isomorphism $d : \mathcal{D}\mathbf{L}^\bullet[n] \cong \mathbf{L}^\bullet$ of an object $\mathbf{L}^\bullet \in SD(X^n)$ to satisfy $\mathcal{D}d[n] = +d$. In the presence of the normalization (**SD1**), this is the correct choice of sign for a space X of *even* dimension: Let $\mathbf{A}^\bullet = \mathbf{L}^\bullet|_{U_2}$ be the restriction to the top stratum. On the manifold U_2 we have

$$\mathcal{D}\mathbf{A}^\bullet[n] \cong R\mathbf{Hom}^\bullet(\mathbf{A}^\bullet, \mathbb{D}^\bullet_{U_2})[n] \cong R\mathbf{Hom}^\bullet(\mathbf{A}^\bullet, or_{U_2}[2n])$$

(or_{U_2} is the orientation sheaf on U_2), so that with $\mathbf{D}(-) = R\mathbf{Hom}^\bullet(-, or_{U_2}[2n])$, the statement $\mathcal{D}d[n] = +d$ means precisely that the diagram

$$\begin{array}{ccc} \mathbf{A}^\bullet & \xrightarrow{d} & \mathbf{DA}^\bullet \\ {\scriptstyle ev}\downarrow & \nearrow{\scriptstyle \mathbf{D}d} & \\ \mathbf{DDA}^\bullet & & \end{array} \tag{29}$$

commutes. Since both $\mathbf{A}^\bullet$ and or_{U_2} are cohomologically locally constant (in fact $\mathbf{A}^\bullet \cong \mathbb{R}_{U_2}[n]$ by (**SD1**)),

$$(\mathbf{DA}^\bullet)_x = R\mathbf{Hom}^\bullet(\mathbf{A}^\bullet_x, or[2n]_x) = \mathrm{Hom}^\bullet(\mathbf{A}^\bullet_x, \mathbb{R}[2n])$$

$(x \in U_2)$. Setting $D(-) = \mathrm{Hom}^\bullet(-, \mathbb{R}[2n])$, diagram (29) is on stalks

$$\text{(30)} \qquad \begin{array}{ccc} \mathbf{A}^\bullet_x & \xrightarrow{d_x} & D\mathbf{A}^\bullet_x \\ {\scriptstyle ev}\downarrow & \nearrow{\scriptstyle Dd_x} & \\ DD\mathbf{A}^\bullet_x & & \end{array}$$

By lemma A.1,

$$\text{(31)} \qquad \begin{array}{ccc} \mathbf{A}^\bullet_x \otimes \mathbf{A}^\bullet_x & \xrightarrow{p(d_x)} & \mathbb{R}[2n] \\ {\scriptstyle sw}\downarrow & \nearrow{\scriptstyle p(d_x)} & \\ \mathbf{A}^\bullet_x \otimes \mathbf{A}^\bullet_x & & \end{array}$$

commutes if and only if (30) commutes. Now $\mathbf{A}^\bullet_x = \mathbb{R}[n]$ and $p^{-2n}(d_x) : \mathbb{R} \otimes \mathbb{R} \to \mathbb{R}$ satisfies $p^{-2n}(d_x)(u \otimes v) = p^{-2n}(d_x)(v \otimes u)$, $u, v \in \mathbb{R}$. Diagram (31) commutes iff

$$p^{-2n}(d_x)(u \otimes v) = p^{-2n}(d_x) \circ sw^{-2n}(u \otimes v) = (-1)^{n^2} p^{-2n}(d_x)(v \otimes u),$$

which is equivalent to n even, as $p(d_x)$ is nonsingular. To work on odd-dimensional spaces, one introduces a category $SD^o(X)$ by changing axiom (**SD4**) to $\mathcal{D}d[n] = -d$ and adapting the rest of the theory accordingly.

Next, we shall discuss the sign of $\mathcal{D}\delta[n+1]$, where $\delta : \mathcal{O}(\mathbf{A}^\bullet) \cong \mathcal{DO}(\mathbf{A}^\bullet)[n+1]$ is the isomorphism given by lemma 2.3 ($\mathbf{A}^\bullet \in SD(U_k)$). Dualizing the distinguished triangle

$$^{\bar{m}}\mathbf{A}^\bullet \xrightarrow{c} {}^{\bar{n}}\mathbf{A}^\bullet \xrightarrow{\omega} \mathcal{O}(\mathbf{A}^\bullet) \xrightarrow{\omega'} {}^{\bar{m}}\mathbf{A}^\bullet[1]$$

yields the distinguished triangle ($\mathcal{O} = \mathcal{O}(\mathbf{A}^\bullet)$)

$$\text{(32)} \qquad \mathcal{DO} \xrightarrow{\mathcal{D}\omega} \mathcal{D}^{\bar{n}}\mathbf{A}^\bullet \xrightarrow{\mathcal{D}c} \mathcal{D}^{\bar{m}}\mathbf{A}^\bullet \xrightarrow{sn \circ \mathcal{D}(\omega'[-1])} (\mathcal{DO})[1],$$

where $sn : \mathcal{D}(\mathcal{O}[-1]) \to (\mathcal{DO})[1]$ is the morphism (28), see [**B**$^+$**84**, V, 5.10], which is however abbreviated as

$$\mathcal{DO} \xrightarrow{\mathcal{D}\omega} \mathcal{D}^{\bar{n}}\mathbf{A}^\bullet \xrightarrow{\mathcal{D}c} \mathcal{D}^{\bar{m}}\mathbf{A}^\bullet \xrightarrow{\mathcal{D}\omega'[1]} (\mathcal{DO})[1]$$

in the main text of this paper. Turning (32) yields

$$\mathcal{D}^{\bar{n}}\mathbf{A}^\bullet \xrightarrow{\mathcal{D}c} \mathcal{D}^{\bar{m}}\mathbf{A}^\bullet \xrightarrow{sn \circ \mathcal{D}(\omega'[-1])} (\mathcal{DO})[1] \xrightarrow{-(\mathcal{D}\omega)[1]} \mathcal{D}^{\bar{n}}\mathbf{A}^\bullet[1]$$

and shifting by $[n]$ we obtain

$$\mathcal{D}^{\bar{n}}\mathbf{A}^\bullet[n] \xrightarrow{\mathcal{D}c[n]} \mathcal{D}^{\bar{m}}\mathbf{A}^\bullet[n] \xrightarrow{sn[n] \circ \mathcal{D}(\omega'[-1])[n]} (\mathcal{DO})[n+1] \xrightarrow{(-1)^{n+1}(\mathcal{D}\omega)[n+1]} \mathcal{D}^{\bar{n}}\mathbf{A}^\bullet[n+1]$$

The identification $\mathcal{D}d[n] = d$ for $d : \mathbf{A}^\bullet \cong \mathcal{D}\mathbf{A}^\bullet[n]$ implies the commutativity of

$$\begin{array}{ccc}
\mathcal{D}^{\bar n}\mathbf{A}^\bullet[n] & \xrightarrow{\mathcal{D}c[n]} & \mathcal{D}^{\bar m}\mathbf{A}^\bullet[n] \\
{\scriptstyle d'}\uparrow{\scriptstyle \cong} & & {\scriptstyle \cong}\uparrow{\scriptstyle \mathcal{D}d'[n]\circ ev} \\
{}^{\bar m}\mathbf{A}^\bullet & \xrightarrow{c} & {}^{\bar n}\mathbf{A}^\bullet
\end{array}$$

where d' is induced by d and $ev : {}^{\bar n}\mathbf{A}^\bullet \longrightarrow \mathcal{D}(\mathcal{D}^{\bar n}\mathbf{A}^\bullet[n])[n]$. The morphism δ completes this square to a morphism of triangles

$$\begin{array}{ccccccc}
\mathcal{D}^{\bar n}\mathbf{A}^\bullet[n] & \xrightarrow{\mathcal{D}c[n]} & \mathcal{D}^{\bar m}\mathbf{A}^\bullet[n] & \xrightarrow{sn[n]\circ\mathcal{D}(\omega'[-1])[n]} & (\mathcal{D}\mathcal{O})[n+1] & \xrightarrow{(-1)^{n+1}(\mathcal{D}\omega)[n+1]} & \mathcal{D}^{\bar n}\mathbf{A}^\bullet[n+1] \\
{\scriptstyle d'}\uparrow{\scriptstyle \cong} & & {\scriptstyle \cong}\uparrow{\scriptstyle \mathcal{D}d'[n]\circ ev} & & {\scriptstyle \cong}\uparrow{\scriptstyle \delta} & & {\scriptstyle \cong}\uparrow{\scriptstyle d'[1]} \\
{}^{\bar m}\mathbf{A}^\bullet & \xrightarrow{c} & {}^{\bar n}\mathbf{A}^\bullet & \xrightarrow{\omega} & \mathcal{O} & \xrightarrow{\omega'} & {}^{\bar m}\mathbf{A}^\bullet[1]
\end{array}$$

Dualizing the square

$$\begin{array}{ccc}
(\mathcal{D}\mathcal{O})[n+1] & \xrightarrow{(-1)^{n+1}(\mathcal{D}\omega)[n+1]} & \mathcal{D}^{\bar n}\mathbf{A}^\bullet[n+1] \\
{\scriptstyle \cong}\uparrow{\scriptstyle \delta} & & {\scriptstyle \cong}\uparrow{\scriptstyle d'[1]} \\
\mathcal{O} & \xrightarrow{\omega'} & {}^{\bar m}\mathbf{A}^\bullet[1]
\end{array}$$

produces the commutative square

$$\begin{array}{ccc}
\mathcal{D}((\mathcal{D}\mathcal{O})[n+1])[n+1] & \xleftarrow{(-1)^{n+1}\mathcal{D}((\mathcal{D}\omega)[n+1])[n+1]} & \mathcal{D}(\mathcal{D}^{\bar n}\mathbf{A}^\bullet[n+1])[n+1] \\
{\scriptstyle \mathcal{D}\delta[n+1]}\downarrow & & \downarrow{\scriptstyle \mathcal{D}(d'[1])[n+1]} \\
(\mathcal{D}\mathcal{O})[n+1] & \xleftarrow{(\mathcal{D}\omega')[n+1]} & \mathcal{D}({}^{\bar m}\mathbf{A}^\bullet[1])[n+1]
\end{array} \tag{33}$$

To determine the sign of $\mathcal{D}\delta[n+1]$, we identify (33) with the commutative square

$$\begin{array}{ccc}
\mathcal{D}^{\bar m}\mathbf{A}^\bullet[n] & \xrightarrow{sn[n]\circ\mathcal{D}(\omega'[-1])[n]} & (\mathcal{D}\mathcal{O})[n+1] \\
{\scriptstyle \cong}\uparrow{\scriptstyle \mathcal{D}d'[n]\circ ev} & & {\scriptstyle \cong}\uparrow{\scriptstyle \delta} \\
{}^{\bar n}\mathbf{A}^\bullet & \xrightarrow{\omega} & \mathcal{O}
\end{array}$$

The identification is achieved via the following four diagrams (34)-(37):

$$\begin{array}{ccc}
\mathcal{D}^{\bar m}\mathbf{A}^\bullet[n] & \xrightarrow{sn[n]\circ\mathcal{D}(\omega'[-1])[n]} & (\mathcal{D}\mathcal{O})[n+1] \\
{\scriptstyle sn[n]}\downarrow{\scriptstyle \cong} & & \Vert \\
\mathcal{D}({}^{\bar m}\mathbf{A}^\bullet[1])[n+1] & \xrightarrow{(\mathcal{D}\omega')[n+1]} & (\mathcal{D}\mathcal{O})[n+1]
\end{array} \tag{34}$$

$$(35) \qquad \begin{array}{ccc} {}^{\bar{n}}\mathbf{A}^\bullet & \xrightarrow[\cong]{ev} & \mathcal{D}(\mathcal{D}^{\bar{n}}\mathbf{A}^\bullet[n+1])[n+1] \\ \Big\downarrow {\scriptstyle \mathcal{D}d'[n]\circ ev} & & \Big\downarrow {\scriptstyle \mathcal{D}(d'[1])[n+1]} \\ \mathcal{D}^{\bar{m}}\mathbf{A}^\bullet[n] & \xrightarrow[\cong]{sn[n]} & \mathcal{D}({}^{\bar{m}}\mathbf{A}^\bullet[1])[n+1] \end{array}$$

$$(36) \qquad \begin{array}{ccc} {}^{\bar{n}}\mathbf{A}^\bullet & \xrightarrow{\qquad\qquad\omega\qquad\qquad} & \mathcal{O} \\ {\scriptstyle ev}\Big\downarrow{\scriptstyle \cong} & & {\scriptstyle \cong}\Big\downarrow{\scriptstyle ev} \\ \mathcal{D}(\mathcal{D}^{\bar{n}}\mathbf{A}^\bullet[n+1])[n+1] & \xrightarrow{(-1)^{n+1}\mathcal{D}((\mathcal{D}\omega)[n+1])[n+1]} & \mathcal{D}((\mathcal{D}\mathcal{O})[n+1])[n+1] \end{array}$$

$$(37) \qquad \begin{array}{ccc} \mathcal{O} & \xrightarrow[\cong]{ev} & \mathcal{D}((\mathcal{D}\mathcal{O})[n+1])[n+1] \\ \Big\downarrow {\scriptstyle \delta} & & \Big\downarrow {\scriptstyle \mathcal{D}\delta[n+1]} \\ \mathcal{D}\mathcal{O}[n+1] & = & \mathcal{D}\mathcal{O}[n+1] \end{array}$$

Straightforward calculations show that (34) commutes and (35) commutes up to $(-1)^{n+1}$. Clearly, (36) commutes up to $(-1)^{n+1}$. Thus, diagram (37) commutes up to $(-1)^{n+1}\,(-1)^{n+1} = +1$, i.e. (37) is commutative.

Let M^{n-k} be the pure stratum $M = U_{k+1} - U_k = X_{n-k} - X_{n-k-1}$, an open manifold of dimension $n-k$. Let $j : M \hookrightarrow U_{k+1}$ denote the (closed) inclusion. Consider the distinguished triangle

$$\begin{array}{ccc} j^!\mathcal{O} & \longrightarrow & j^*\mathcal{O} \\ {\scriptstyle [1]}\nwarrow & & \swarrow \\ & j^*Ri_{k*}i_k^*\mathcal{O} & \end{array}$$

in $D^b(M)$. In $D^b(U_k)$, $i_k^*\mathcal{O} \cong \mathbf{0}^\bullet$ (the zero complex), since $\mathcal{O}$ is supported in M. Thus the canonical morphism $j^!\mathcal{O} \to j^*\mathcal{O}$ is an isomorphism in $D^b(M)$. Consequently, $j^*\mathcal{D}\mathcal{O}[n+1] = \mathcal{D}(j^!\mathcal{O})[n+1] = \mathcal{D}(j^*\mathcal{O})[n+1]$. Thus, applying j^* to diagram (37), we obtain

$$(38) \qquad \begin{array}{ccc} j^*\mathcal{O} & \xrightarrow{j^*\delta} & \mathcal{D}(j^*\mathcal{O})[n+1] \\ \Big\downarrow {\scriptstyle ev} & \nearrow {\scriptstyle \mathcal{D}(j^*\delta)[n+1]} & \\ \mathcal{D}(\mathcal{D}(j^*\mathcal{O})[n+1])[n+1] & & \end{array}$$

Set $\mathcal{S} = j^*\mathcal{O}$, $\zeta = j^*\delta$ and let $x \in M$. On a manifold, the dualizing complex is canonically isomorphic to the shifted orientation sheaf,

$$\mathbb{D}_M^\bullet = or_M[n-k],$$

and hence

$$\mathcal{D}\mathcal{S}[n+1] = R\mathbf{Hom}^\bullet(\mathcal{S}, \mathbb{D}^\bullet_M[n+1]) = R\mathbf{Hom}^\bullet(\mathcal{S}, or_M[t]),$$

with $t = 2n - k + 1$. Since both $\mathcal{S}$ and $or_M[t]$ are cohomologically locally constant on M, we have

$$R\mathbf{Hom}^\bullet(\mathcal{S}, or_M[t])_x = R\,\mathrm{Hom}^\bullet(\mathcal{S}_x, or_M[t]_x) = \mathrm{Hom}^\bullet(\mathcal{S}_x, \mathbb{R}[t]).$$

Thus, diagram (38) restricts at x to the commutative diagram

$$\begin{array}{ccc} \mathcal{S}_x & \xrightarrow{\zeta_x} & \mathrm{Hom}^\bullet(\mathcal{S}_x, \mathbb{R}[t]) \\ {\scriptstyle ev}\downarrow & \nearrow {\scriptstyle \mathrm{Hom}^\bullet(\zeta_x, \mathbb{R}[t])} & \\ \mathrm{Hom}^\bullet(\mathrm{Hom}^\bullet(\mathcal{S}_x, \mathbb{R}[t]), \mathbb{R}[t]) & & \end{array}$$

By lemma A.1, the induced diagram

$$\begin{array}{ccc} \mathcal{S}_x \otimes \mathcal{S}_x & \xrightarrow{p(\zeta_x)} & \mathbb{R}[t] \\ {\scriptstyle sw}\downarrow & \nearrow {\scriptstyle p(\zeta_x)} & \\ \mathcal{S}_x \otimes \mathcal{S}_x & & \end{array}$$

commutes. Now k is odd, $k = 2l + 1$. Note that then l is the middle dimension of the links of the components of M. The complex $\mathcal{S}_x$ is concentrated in degree $s = \bar{n}(k) - n = l - n$, and $2s = -t$. For $u, v \in \mathcal{S}^s_x$ we compute

$$\begin{aligned} p^{2s}(\zeta_x)(u \otimes v) &= p^{2s}(\zeta_x) \circ sw^{2s}(u \otimes v) \\ &= (-1)^{s^2} p^{2s}(\zeta_x)(v \otimes u) \\ &= (-1)^{l-n} p^{2s}(\zeta_x)(v \otimes u) \\ &= (-1)^{l} p^{2s}(\zeta_x)(v \otimes u), \end{aligned}$$

as n is even if $\mathbf{A}^\bullet \in SD(U_k)$. This proves that the pairing induced by δ is symmetric if the dimension of the link is $\equiv 0(4)$ and anti-symmetric if the dimension is $\equiv 2(4)$.

Bibliography

[B$^+$84] A. Borel et al., *Intersection cohomology*, Progr. Math., no. 50, Birkhäuser Verlag, Boston, 1984.

[Ban99] M. Banagl, *Extending intersection homology type invariants to non-Witt spaces*, Ph.D. thesis, New York University, 1999.

[BC00] M. Banagl and S. E. Cappell, *Computing twisted signatures and L-classes of stratified spaces*, Preprint (2000).

[BK01] M. Banagl and R. Kulkarni, *Self-dual sheaves on the reductive Borel-Serre compactification of Hilbert modular surfaces*, in preparation (2001).

[Che79] J. Cheeger, *On the spectral geometry of spaces with cone-like singularities*, Proc. Natl. Acad. Sci. USA **76** (1979), 2103 – 2106.

[Che80] J. Cheeger, *On the Hodge theory of Riemannian pseudomanifolds*, Proc. Sympos. Pure Math. **36** (1980), 91–146.

[Che83] J. Cheeger, *Spectral geometry of singular Riemannian spaces*, J. Differential Geom. **18** (1983), 575 – 657.

[Coh67] M. Cohen, *Simplicial structures and transverse cellularity*, Ann. of Math. **85** (1967), 218 – 245.

[CS91] S. E. Cappell and J. L. Shaneson, *Stratifiable maps and topological invariants*, J. Amer. Math. Soc. **4** (1991), 521–551.

[CSW91] S. E. Cappell, J. L. Shaneson, and S. Weinberger, *Classes topologiques caractéristiques pour les actions de groupes sur les espaces singuliers*, C. R. Acad. Sci. Paris Sér. I Math. **313** (1991), 293–295.

[GHM94] M. Goresky, G. Harder, and R. D. MacPherson, *Weighted cohomology*, Invent. Math. **116** (1994), 139 – 213.

[GM80] M. Goresky and R. D. MacPherson, *Intersection homology theory*, Topology **19** (1980), 135 – 162.

[GM83] M. Goresky and R. D. MacPherson, *Intersection homology II*, Invent. Math. **71** (1983), 77 – 129.

[Ive86] Birger Iversen, *Cohomology of sheaves*, Universitext, Springer Verlag, 1986.

[KS90] Masaki Kashiwara and Pierre Schapira, *Sheaves on manifolds*, Grundlehren Math. Wiss., no. 292, Springer Verlag, 1990.

[McC75] C. McCrory, *Cone complexes and PL–transversality*, Trans. Amer. Math. Soc. **207** (1975), 1–23.

[Qui88] F. Quinn, *Homotopically stratified sets*, J. Amer. Math. Soc. **1** (1988), 441 – 499.

[RS68] C. P. Rourke and B. J. Sanderson, *Block bundles I*, Ann. of Math. **87** (1968), 1 – 28.

[Sie83] P. H. Siegel, *Witt spaces: A geometric cycle theory for KO-homology at odd primes*, Amer. J. Math. **105** (1983), 1067–1105.

[Sto68] R. E. Stong, *Notes on cobordism theory*, Math. Notes, Princeton University Press, 1968.

[Sto72] D. A. Stone, *Stratified polyhedra*, Lecture Notes in Math., no. 252, Springer Verlag, New York, 1972.

[Ver84] Andrei Verona, *Stratified mappings – structure and triangulability*, Lecture Notes in Math., no. 1102, Springer Verlag, New York, 1984.

[Wal99] C. T. C. Wall, *Surgery on compact manifolds*, second ed., Mathematical Surveys and Monographs, vol. 69, American Mathematical Society, 1999.

[Wei94] S. Weinberger, *The topological classification of stratified spaces*, Chicago Lectures in Math., Univ. of Chicago Press, Chicago, 1994.

Editorial Information

To be published in the *Memoirs*, a paper must be correct, new, nontrivial, and significant. Further, it must be well written and of interest to a substantial number of mathematicians. Piecemeal results, such as an inconclusive step toward an unproved major theorem or a minor variation on a known result, are in general not acceptable for publication. Papers appearing in *Memoirs* are generally longer than those appearing in *Transactions*, which shares the same editorial committee.

As of June 30, 2002, the backlog for this journal was approximately 5 volumes. This estimate is the result of dividing the number of manuscripts for this journal in the Providence office that have not yet gone to the printer on the above date by the average number of monographs per volume over the previous twelve months, reduced by the number of volumes published in four months (the time necessary for preparing a volume for the printer). (There are 6 volumes per year, each containing at least 4 numbers.)

A Consent to Publish and Copyright Agreement is required before a paper will be published in the *Memoirs*. After a paper is accepted for publication, the Providence office will send a Consent to Publish and Copyright Agreement to all authors of the paper. By submitting a paper to the *Memoirs*, authors certify that the results have not been submitted to nor are they under consideration for publication by another journal, conference proceedings, or similar publication.

Information for Authors

Memoirs are printed from camera copy fully prepared by the author. This means that the finished book will look exactly like the copy submitted.

The paper must contain a *descriptive title* and an *abstract* that summarizes the article in language suitable for workers in the general field (algebra, analysis, etc.). The *descriptive title* should be short, but informative; useless or vague phrases such as "some remarks about" or "concerning" should be avoided. The *abstract* should be at least one complete sentence, and at most 300 words. Included with the footnotes to the paper should be the 2000 *Mathematics Subject Classification* representing the primary and secondary subjects of the article. The classifications are accessible from `www.ams.org/msc/`. The list of classifications is also available in print starting with the 1999 annual index of *Mathematical Reviews*. The Mathematics Subject Classification footnote may be followed by a list of *key words and phrases* describing the subject matter of the article and taken from it. Journal abbreviations used in bibliographies are listed in the latest *Mathematical Reviews* annual index. The series abbreviations are also accessible from `www.ams.org/publications/`. To help in preparing and verifying references, the AMS offers MR Lookup, a Reference Tool for Linking, at `www.ams.org/mrlookup/`. When the manuscript is submitted, authors should supply the editor with electronic addresses if available. These will be printed after the postal address at the end of the article.

Electronically prepared manuscripts. The AMS encourages electronically prepared manuscripts, with a strong preference for AMS-LaTeX. To this end, the Society has prepared AMS-LaTeX author packages for each AMS publication. Author packages include instructions for preparing electronic manuscripts, the *AMS Author Handbook*, samples, and a style file that generates the particular design specifications of that publication series. Though AMS-LaTeX is the highly preferred format of TeX, author packages are also available in AMS-TeX.

Authors may retrieve an author package from e-MATH starting from www.ams.org/tex/ or via FTP to ftp.ams.org (login as anonymous, enter username as password, and type cd pub/author-info). The *AMS Author Handbook* and the *Instruction Manual* are available in PDF format following the author packages link from www.ams.org/tex/. The author package can be obtained free of charge by sending email to pub@ams.org (Internet) or from the Publication Division, American Mathematical Society, P.O. Box 6248, Providence, RI 02940-6248. When requesting an author package, please specify AMS-LaTeX or AMS-TeX, Macintosh or IBM (3.5) format, and the publication in which your paper will appear. Please be sure to include your complete mailing address.

Sending electronic files. After acceptance, the source file(s) should be sent to the Providence office (this includes any TeX source file, any graphics files, and the DVI or PostScript file).

Before sending the source file, be sure you have proofread your paper carefully. The files you send must be the EXACT files used to generate the proof copy that was accepted for publication. For all publications, authors are required to send a printed copy of their paper, which exactly matches the copy approved for publication, along with any graphics that will appear in the paper.

TeX files may be submitted by email, FTP, or on diskette. The DVI file(s) and PostScript files should be submitted only by FTP or on diskette unless they are encoded properly to submit through email. (DVI files are binary and PostScript files tend to be very large.)

Electronically prepared manuscripts can be sent via email to pub-submit@ams.org (Internet). The subject line of the message should include the publication code to identify it as a Memoir. TeX source files, DVI files, and PostScript files can be transferred over the Internet by FTP to the Internet node e-math.ams.org (130.44.1.100).

Electronic graphics. Comprehensive instructions on preparing graphics are available at www.ams.org/jourhtml/graphics.html. A few of the major requirements are given here.

Submit files for graphics as EPS (Encapsulated PostScript) files. This includes graphics originated via a graphics application as well as scanned photographs or other computer-generated images. If this is not possible, TIFF files are acceptable as long as they can be opened in Adobe Photoshop or Illustrator. No matter what method was used to produce the graphic, it is necessary to provide a paper copy to the AMS.

Authors using graphics packages for the creation of electronic art should also avoid the use of any lines thinner than 0.5 points in width. Many graphics packages allow the user to specify a "hairline" for a very thin line. Hairlines often look acceptable when proofed on a typical laser printer. However, when produced on a high-resolution laser imagesetter, hairlines become nearly invisible and will be lost entirely in the final printing process.

Screens should be set to values between 15% and 85%. Screens which fall outside of this range are too light or too dark to print correctly. Variations of screens within a graphic should be no less than 10%.

Inquiries. Any inquiries concerning a paper that has been accepted for publication should be sent directly to the Electronic Prepress Department, American Mathematical Society, P. O. Box 6248, Providence, RI 02940-6248.

Editors

732 **Jan O. Kleppe, Juan C. Migliore, Rosa Miró-Roig, Uwe Nagel, and Chris Peterson,** Gorenstein liaison, complete intersection liaison invariants and unobstructedness, 2001

731 **Jesús Bastero, Mario Milman, and Francisco J. Ruiz,** On the connection between weighted norm inequalities, commutators and real interpolation, 2001

730 **Suhyoung Choi,** The decomposition and classification of radiant affine 3-manifolds, 2001

729 **Michael Grosser, Eva Farkas, Michael Kunzinger, and Roland Steinbauer,** On the foundations of nonlinear generalized functions I and II, 2001

728 **Laura Smithies,** Equivariant analytic localization of group representations, 2001

727 **Anthony D. Blaom,** A geometric setting for Hamiltonian perturbation theory, 2001

726 **Victor L. Shapiro,** Singular quasilinearity and higher eigenvalues, 2001

725 **Jean-Pierre Rosay and Edgar Lee Stout,** Strong boundary values, analytic functionals, and nonlinear Paley-Wiener theory, 2001

724 **Lisa Carbone,** Non-uniform lattices on uniform trees, 2001

723 **Deborah M. King and John B. Strantzen,** Maximum entropy of cycles of even period, 2001

722 **Hernán Cendra, Jerrold E. Marsden, and Tudor S. Ratiu,** Lagrangian reduction by stages, 2001

721 **Ingrid C. Bauer,** Surfaces with $K^2 = 7$ and $p_g = 4$, 2001

720 **Palle E. T. Jorgensen,** Ruelle operators: Functions which are harmonic with respect to a transfer operator, 2001

719 **Steve Hofmann and John L. Lewis,** The Dirichlet problem for parabolic operators with singular drift terms, 2001

718 **Bernhard Lani-Wayda,** Wandering solutions of delay equations with sine-like feedback, 2001

717 **Ron Brown,** Frobenius groups and classical maximal orders, 2001

716 **John H. Palmieri,** Stable homotopy over the Steenrod algebra, 2001

715 **W. N. Everitt and L. Markus,** Multi-interval linear ordinary boundary value problems and complex symplectic algebra, 2001

714 **Earl Berkson, Jean Bourgain, and Aleksander Pełczynski,** Canonical Sobolev projections of weak type $(1,1)$, 2001

713 **Dorina Mitrea, Marius Mitrea, and Michael Taylor,** Layer potentials, the Hodge Laplacian, and global boundary problems in nonsmooth Riemannian manifolds, 2001

712 **Raúl E. Curto and Woo Young Lee,** Joint hyponormality of Toeplitz pairs, 2001

711 **V. G. Kac, C. Martinez, and E. Zelmanov,** Graded simple Jordan superalgebras of growth one, 2001

710 **Brian Marcus and Selim Tuncel,** Resolving Markov chains onto Bernoulli shifts via positive polynomials, 2001

709 **B. V. Rajarama Bhat,** Cocylces of CCR flows, 2001

708 **William M. Kantor and Ákos Seress,** Black box classical groups, 2001

707 **Henning Krause,** The spectrum of a module category, 2001

706 **Jonathan Brundan, Richard Dipper, and Alexander Kleshchev,** Quantum Linear groups and representations of $GL_n(\mathbb{F}_q)$, 2001

705 **I. Moerdijk and J. J. C. Vermeulen,** Proper maps of toposes, 2000

704 **Jeff Hooper, Victor Snaith, and Min van Tran,** The second Chinburg conjecture for quaternion fields, 2000

For a complete list of titles in this series, visit the AMS Bookstore at **www.ams.org/bookstore/**.